Identity Politics and Elections
in Malaysia and I

T0259049

In recent social research, ethnicity has mostly been used as an explanatory variable. Only after it was agreed that ethnicity, in itself, is subject to change was it possible to answer the questions of how and why it changes. This multiplicity of ethnic identities requires that we think of each society as one with multiple ethnic dimensions, of which any can become activated in the process of political competition – and sometimes several of them within a short period of time.

Focusing on Malaysia and Indonesia, this book traces the variations of ethnic identity by looking at electoral strategies in two sub-national units. It shows that ethnic identities are subject to change – induced by calculated moves by political entrepreneurs who use identities as tools to maximize their chances of winning elections or expanding support base – and highlights how political institutions play an enormous role in shaping the modes and dynamics of these ethno-political manipulations. This book suggests that in societies where ethnic identities are activated in politics, instead of analyzing politics with ethnic distribution as an independent variable, ethnic distribution can be taken as the dependent variable, with political institutions being the explanatory one. It examines the problems of voters' behaviour, and parties' and candidates' strategy in a polity that is, to a significant extent, driven by ethnic relations.

Pushing the boundaries of qualitative research on Southeast Asian politics by placing formal institutions at the centre of its analysis, this book will be of interest to students and scholars of southeast Asian politics, race and ethnic studies and international relations.

Karolina Prasad gained a PhD from Hamburg University, Germany in 2013, after carrying out extensive research in Borneo. She also graduated with an MA in political science from Warsaw University, Poland, in 2006.

Routledge Contemporary Southeast Asia Series

Identity Politics and Elections in Malaysia and Indonesia

Ethnic engineering in Borneo

Karolina Prasad

LONDON AND NEW YORK

First published 2016 by Routledge

2 Park Square, Milton Park, Abingdon, Oxfordshire OX14 4RN
711 Third Avenue, New York, NY 10017

Routledge is an imprint of the Taylor & Francis Group, an Informa business

First issued in paperback 2018

British Library Cataloguing in Publication Data
A catalogue record for this book is available from the British Library

Library of Congress Cataloging-in-Publication Data
Names: Prasad, Karolina, author.
Title: Identity politics and elections in Malaysia and Indonesia : ethnic
    engineering in Borneo / Karolina Prasad.
Description: New York : Routledge, 2016. | Series: Routledge
    contemporary Southeast Asia series ; 105 | Includes bibliographical
    references and index.
Identifiers: LCCN 2015024469 | ISBN 9781138854734 (hardback) |
    ISBN 9781315720876 (ebook)
Subjects: LCSH: Sarawak (Malaysia)—Ethnic relations—Political aspects. |
    Kalimantan Barat (Indonesia)—Ethnic relations—Political aspects. |
    Identity politics—Malaysia—Sarawak. | Identity politics—Indonesia—
    Kalimantan Barat. | Elections—Malaysia—Sarawak. | Elections—
    Indonesia—Kalimantan Barat.
Classification: LCC DS597.366 .P73 2016 | DDC 305.809595/4—dc23
LC record available at http://lccn.loc.gov/2015024469

ISBN: 978-1-138-85473-4 (hbk)
ISBN: 978-1-138-61108-5 (pbk)

Typeset in Times New Roman
by Apex CoVantage, LLC

Contents

Figures

Tables

1   Theories, institutions and ethnicity

1.1 Ethnicity and politics

For most of the 20th century, two major themes occupied students of ethno-politics: the first was the theoretical debate on the nature of ethnicity itself and the contents of the concept. The second theme was the practical question of how to design state institutions to eliminate the presumably destructive forces of ethnic diversity. The former problem, typically presented in the literature as a dispute between "primordialists", who see ethnic identity as single and fixed, and "constructivists", to whom ethnic identity is multidimensional and fluid, was finally declared as solved by Brubaker, who pronounced primordialism "a long dead horse" (1996, 15).[1] The latter issue of institutional design has been dealt with in a plethora of empirical studies, both qualitative and quantitative, but the results so far have been less than satisfactory, and there is no clear answer to the question of what institutional setting could ultimately minimize the chance of occurrence of conflict in plural societies. The mere assumption that multiethnic states are more prone to conflicts and politically less stable than (relatively) homogeneous states (Horowitz 1985; Rabushka and Shepsle 2009) rests on a premise that is in discordance with the newest findings of empirical studies (Brubaker and Laitin 1998; Davidson 2008a).

Despite the first theme seemingly being closed[2] and the second hardly lending itself to any hope of solution, this work nevertheless attempts to contribute to the two debates, and does it for two reasons. Foremost, eliminating primordialist elements from the ethno-theoretic discourse did not mean that the constructivist camp agreed on at least the most basic properties and dynamics of change of ethnic identities. Some new theoretical propositions that have arisen from the constructivist triumph over primordialism are promising but they are yet to be tested in confrontation with real-life phenomena. Therefore, the explanatory power of Rogers Brubaker's (1996, 2004) and Kanchan Chandra's (2004, 2008, 2012b) concepts, as two of the potentially most powerful conceptual tools in ethnic studies, will be assessed here.

Because, as Brubaker's argument goes, "ethnic groups" are neither "substantial entities to which interests and agency can be attributed" nor "unitary collective actors with common purposes" (2002, 163), "group" is hardly a useful concept for political analysts. Brubaker's work has liberated the discussion on identity from the iron cage of "groups" and shifted the analysis to the level of an individual, which paved the way for Chandra's concept of "ethnic categories". Therefore, we have arrived at the point where there is no one ethnic identity for each individual, and there is no one ethnic group to which she belongs and acts accordingly – multiple categories of identities exist for each individual, which are invoked in daily life and political behaviour. While this line of thinking is not new (compare Barth 1970), an analytical approach that lends the paradigm to operationalization and comparable studies has recently been systematized and is worth exploring.

The second underlying reason for this study is the consequence of this general agreement that ethnic identity can and does change along with changing incentives. Here the "change" refers to a shift from one category, with which a person identifies (say, "Northerner"), to another category, in which she is also a member (e.g. "Christian"), and at some other time or circumstance to yet another ("Black"), while nominally being a Black Christian Northerner.[3] All these categories (and likely a few others) constitute this person's ethnic repertoire.[4] So the compelling question is why and how often people shift from one category to another within their repertoire, which in politics translates into a question: how do institutions induce the change and impact its frequency? Consequently, what institutions should states deploy if they face a multiethnic demography? Institutional prescriptions for plural societies developed in the 20th century, while nominally acknowledging the fluidity of ethnic identity, did not take into account the actual possibility of change. Ethnic distribution of analyzed societies was seen as fixed and relied on one arbitrarily chosen (or census-established[5]) dimension, which led authors to provide scientifically developed solutions to the conflict potential in a static, single-dimension polity.

The two states analyzed in this research correspond to the two paradigms of institutional design for divided societies: consociationalism proposed by Arend Lijphart (1969, 2004), and centripetalism[6] proposed by Donald Horowitz (1985, 1991) and systematized by Benjamin Reilly (2011). One important question is whether these existing recommendations for plural societies succeed in maintaining peaceful and democratic polities, and how they achieve that goal – what are the actual, day-to-day mechanisms of political behaviour that induce non-conflict modes of coexistence between ethnic categories? These designs have a strong standing in political theory and practice and are a ready set of hypotheses and assumptions about patterns of ethnic dynamics in societies in which they function. States frequently deploy propositions embedded in these designs, and comparing ethnic identity change may be one way of assessing how these institutions affect the inter-ethnic relations in these states.

Consociationalism, proposed by Arend Lijphart, in simplest terms is based on power sharing, or "the participation of representatives of all significant communal groups in political decision making, especially at the executive level; [and] group autonomy [which] means that these groups have authority to run their own internal affairs, especially in the areas of education and culture" (Lijphart 2004, 97).[7] Consociationalism implies that it is institutionally prescribed with which category in their repertoire people should identify, and usually contains all or most of the following: grand coalition governments with all ethnic components deemed relevant involved; proportional representation of different groups in the legislative and in the civil service; segmental autonomy; and a power of veto in case of decisions crucial to ethnic components (Reilly 2006, 815).

Electoral arrangements consociationalists advocate include proportional voting systems and ethnically based parties, and also require "elite-level negotiations between the leaders of the various groups" (Reilly 2006, 815). However, the consociational set-up can involve a variety of different institutional regulations, as Lijphart pointed out in this enumeration of real-life applications:

> Broad representation in the executive has been achieved by a constitutional requirement that it be composed of equal numbers of the two major ethnolinguistic groups (Belgium); by granting all parties with a minimum of 5 percent of the legislative seats the right to be represented in the cabinet (South Africa, 1994–99); by the equal representation of the two main parties in the cabinet and an alternation between the two parties in the presidency (Colombia, 1958–64); and by permanently earmarking the presidency for one group and the prime ministership for another (Lebanon).
>
> (2004, 99)

The opposite of consociationalism is centripetalism, which "eschews the reification of ethnic identity inherent in consociationalism and communalism, instead advocating the need for aggregative, centrist and inter-ethnic politics in divided societies" (Reilly 2011, 4). Centripetal-oriented institutions aim to "dilute the ethnic character of competitive politics and promote multiethnic outcomes instead. This means that, for instance, rather than focusing on the fair representation of ethnically-defined political parties, centripetalists place a premium on promoting multiethnic parties and cross-ethnic activity instead" (Reilly 2011, 5). Several institutional incentives were identified for creating multiethnic bridging and political mobilization outside of one's category. For instance, the goal can be achieved by imposing a ban on ethnic parties, or by the requirement that to win election a presidential candidate needs to obtain his support from not only the majority of voters, but also from voters spread across different ethnic components (usually expressed in geographic terms).

The two cases analyzed in this book correspond to the rationales informing the two approaches political science proposes to the problem of divided societies, although none of them is a perfect realization of the two systems. Malaysia (and Sarawak within it) fulfils several criteria of consociational democracy; power is shared between predefined ethnic categories within a grand coalition and each citizen can claim membership in at least one of the participating ethnic categories; the representation of the ethnic components in the legislative is divided more or less proportionally, although the proportionality is not a result of proportional elections (generally advocated by consociationalism), but results from assigning constituencies/seats to each ethnic category. Some autonomy in cultural affairs is granted to ethnic categories,[8] and the element of elite negotiations is very strong.[9]

Consociational design should be expected to reify and arrest ethnic identity for extended periods of time: power is shared and negotiated between the same categories over and over again (e.g. between Muslims and Christians, Flemish and French speakers, Blacks and Whites), and there is little space for activation of other categories. Sarawak in Malaysia, therefore, is used here as a case study to test whether and how ethnic categories are indeed arrested in a consociational polity, or if the ethnic identity change still happens in this society, thanks to the institutions in which it takes place.

Indonesia and one of its provinces selected for this study, West Kalimantan, broadly correspond to the centripetalist design. Along with direct presidential elections in 2004, the country introduced a stipulation for winning presidential candidates not only to obtain the majority of votes, but also to gain support from at least 20% of voters in at least half of all the provinces in Indonesia, along with the ban of regional parties (which as a result also bans regional ethnic parties), a proportional electoral system and incentives for parties to build coalitions to nominate executive candidates. Ethnic mobilization is frowned upon with one important exception – religion, specifically Islam. Parties calling for more Islamic influence in the state operate freely and individual candidates (also from non-Islamic parties) involve religious elements in their campaigns. This way, Indonesia fulfils most – but again, not all – of the criteria to be recognized as a centripetalist design. Although centripetalist institutions are designed to divert political discourse away from ethnically driven interests, this has not been so far achieved in Indonesia; politics is very much focussed around ethnicity, all the more so at the sub-national level. The search for a mechanism of politicization of ethnicity within a seemingly centripetal setting is one of the tasks of this research.

In an effort to overcome the centripetalism versus consociationalism debate, Kanchan Chandra followed through with her theoretical propositions and, assuming multiple and fluid ethnic identities for each individual, she put forward a pragmatic argument: fluidity of ethnic identities offers a potential for an alternative institutional design for plural societies. Chandra suggests that the key is to induce such a combination of institutions that leads an individual to frequently shift between her identity categories. Simply put, institutions should be such that they neither prescribe identifying with a particular category (as it is in consociationalism), nor should they get in the way of ethnic mobilization (2008, 27). Chandra named the process an "ethnic invention", as it encourages political activation of ethnic categories that otherwise might remain only nominal and "create[s] incentives for voters to retain multiple identities in their repertoires" (2008, 27). Moreover, "ethnic invention" is conditional on lack of constraints for ethnic mobilization in a polity (e.g. ethnic party bans).

Simultaneously, therefore, Malaysia and Indonesia will be test cases for the question hidden behind Chandra's (2008) proposition of an alternative institutional approach for divided societies. Chandra's proposition implies that each election can be and should be an opportunity for a voter to identify with a different identity from her ethnic repertoire, and hence, the more elections in which the voter participates, the more activated (i.e. politically relevant) identities she retains in her repertoire, and the more stable the polity is. Malaysia and Indonesia are excellent case studies to test this hypothesis: a state with few elected offices (Malaysia holds a total of only two elections: to the state and federal legislatures), and a state with a total of seven elections for different offices (Indonesia has direct executive and legislative elections on three tiers) will be compared to see in which of them voters retain more categories in their ethnic repertoires. As we saw earlier, however, Malaysia places no obstacles on ethnic mobilization, which makes ethnic mobilization easier, while Indonesia limits opportunities for ethnic mobilization through its centripetal institutions. We will ask: in which of the two studied cases is a more frequent and faster shift between identity categories observed? And to which particular elements of the institutional maze should the identity shift (and lack thereof) be attributed?

Being committed to analysis of ethnic identity change, we do wish to ask: what are the ways of inducing or hampering ethnic identity shifts under these three institutional designs? The three designs have specific goals, which, if achieved, would lead to different outcomes. Consociationalism wants to achieve democratic stability through the long-term arrest of ethnic categories (under the implications of this system each individual is induced to identify with only one category over long periods of time and in all political circumstances). Centripetalism, on the other hand, attempts to make all ethnic categories politically irrelevant as it discourages mobilizing of any ethnic categories. Chandra's "ethnic inventionism" seeks to disperse the ethnic loyalty of each individual between several categories. By addressing these issues, we will arrive at some interesting conclusions regarding the stability of political and social life under the different settings. West Kalimantan has experienced several violent episodes, including recent and deadly ones, and this experience is assumed to factor in day-to-day political proceedings. Sarawak has remained peaceful over the decades despite the 1969 turmoil in Peninsular Malaysia, and we need to ask what contributed to Sarawak's social stability.

Let it be underscored that the distinction here between consociationalism, centripetalism and "ethnic inventionism" serves the purpose of establishing the research design, as it offers a coherent theoretical approach to the question of institutions and ethnicity and helps generate hypotheses. Although these institutional designs offer a ready-made set of 0, 1 value variables and differentiate neatly between the cases, the actual analysis of this work will look into each element of the institutional system as a distinct variable. A close observation of two cases over a period of time will allow capturing the impact of each of the institutional elements on ethnic identity change. Hypotheses based on the presented institutional designs for divided societies are presented further in this chapter.

1.2 Malaysia and Sarawak, Indonesia and West Kalimantan – how to compare?

When it came to my attention several years ago that the island of Borneo was shared by three different state entities, Malaysia, Indonesia and Brunei, I did not immediately recognize the academic potential that lies in this historic-political setting. I set off to investigate the nexus of ethnicity and politics in what I identi-fied as the most similar regions on Borneo, Sarawak in Malaysia and West Kali-mantan in Indonesia. The potential of comparison lies in the fact, I concluded, that the *ethnic structures* – which should be understood as "set of identities that are considered commonsensically real by a population, whether or not individuals actually identify with them" (Chandra 2009, 4) – of the two regions are almost identical, while the *ethnic practices*, i.e. the activated ethnic categories, or these categories with which people actually identify, in the two cases are substantially different, and the question appeared: why did people identify with different cat-egories across the border if the contents of their ethnic repertoires were the same? Were political institutions behind these identity choices?

Local academic and popular interpretations of politics looked at ethnicity (understood in simple vernacular terms) from a different perspective: as the key to explain politics. The existing body of literature on local politics – for Sarawak quite rich indeed – invariably offered interpretations of politics through the prism of ethnic relations (Chin 1996; Hazis 2012; Jayum 1991; Leigh 1974; Searle 1983). Albeit ethnographically insightful, most of these interpretations rested on the primordial understanding of ethnicity: ethnic identities are fixed. Analyses of electoral results based on these primordial assumptions commonly seek to find correlation of the voters' ethnic identity and that of the elected candidate (which they indeed find) and so the conclusion reads: people vote for their co-ethnics. The explanatory power of this argument is – after a closer look at the political scene of the analyzed cases – minimal. In the case of Malaysia, which is designed as a consociational polity where the elites strike an ethnic bargain to share the political power, almost all constituencies are habitually identified by their ethnic majority (e.g. "Chinese seat" or "Malay seat"). Consequently, candidates in a given con-stituency are of the same ethnic background as their voters. A 90% Chinese–10% Malay voting district always sees a battle between two Chinese candidates, who most likely speak the same Chinese dialect and follow the same religion, so no preference on the candidates' ethnicity is expressed through the ballot paper. We only rarely observe inter-ethnic competition in Malaysia; even in ethnically mixed districts, we usually witness ethnically homogeneous candidates. This is not to say that ethnicity does not matter. Ethnic arguments are often prominently raised during candidate selection and outside the electoral context, e.g. on the occasion of ministerial nominations, and ethnic undertones are by no means absent from electoral campaigns. Nevertheless, to conclude from an election in Malaysia that ethnicity is *the* relevant political factor here is logically and methodologically troublesome, and suggests that a change of perspective is necessary in order to shed some new light on ethno-politics in Malaysia.

A study of the electoral outcomes in Indonesia offers equally puzzling results, especially if one would like to learn about the role of ethnicity for this polity. Within this proportional electoral system tracing ethnicity is almost impossible, unless inferred from party support. At the same time, confronted with the direct executive elections we observe, on one hand, that most candidates' support is split exactly along the religious lines; on the other hand, however, in many districts the combined ticket (i.e. governor and vice-governor) is split between candidates of different ethnic backgrounds and, indeed, in West Kalimantan the winning coalition for the 2007 and 2012 gubernatorial elections was "Dayak (a local ethnic category) and Chinese", while several Malay-Dayak winning teams won on the district level. The compromise and inter-ethnic negotiations therefore do exist and ethnic deals are struck, despite the common assumption of unavoidability of conflict or "incompatibility" of certain ethnic categories (e.g. "Muslim" and "Christian").

Therefore, ethno-political explanations, so popular in the vernacular discourse, media reports and academic research, that involve rigid categorization of the society, e.g. Malays-Dayaks-Chinese, fail to represent the full scope of ethnicity in politics. Shortcomings of these discourses usually lie in taking "ethnic groups" for granted: "Malay" or "Dayak" are used as self-explanatory categories. They are defined by their religious, linguistic or cultural properties, are supposed to share some common history, presumably have coherent political views and, bottom line, are a single collective actor. While we discard this primordial understanding of these categories, we should nevertheless recognize these vernacular categorizations as important pieces of information. Popular usage of certain ethnic labels is a good start of the search for politically activated categories.

Both regions analyzed in this study are provincial and remote from their respective "centres" – Kuala Lumpur and Jakarta – not only geographically, but also in terms of ethnic composition and local power distribution. Sarawak (along with Sabah) is legally distinguished from West Malaysia: a glimpse into the Malaysian constitution, history textbooks, statistical yearbooks and electoral results will provide enough arguments to essentialize Sarawak. Indeed, social scientists habitually separate the Bornean states of Malaysia from the Malay Peninsula and either limit their studies to one of the three, or provide different explanations for each. West Kalimantan, on the other hand, is in many ways just one of the 34 provinces in Indonesia. West Kalimantan's distinctiveness lies in the fact that it is one of only seven provinces[10] in which Islam is professed by less than 60% of residents (Na'im and Syaputra 2012, 11); West Kalimantan among all provinces also has the highest percentage (6.6) of residents who in their daily communication use a "foreign language" (*bahasa asing*), distinguished from "local languages" (Na'im and Syaputra 2012, 13). It has to be assumed that the "foreign language" refers to some Chinese dialects used by the Indonesians of Chinese origin.[11] Although the province's ethnic structure makes it quite particular, no institutional provisions exist to accommodate this specificity. However, as in the case of Malaysia, it would be difficult to conclude anything about Indonesian politics as a whole that would equally refer to West Kalimantan.

What is the axis of comparison of the two units? Given the impossibility of fulfilling criteria for experimental studies in social research, and having even *quasi-experiments* declared misleading (Collier, Brady and Seawright 2004, 231), social sciences have to settle for observational studies, with all weaknesses and traps nested in them (Collier, Brady and Seawright 2004, 230–233). The cases analyzed here – sub-national units of states – were not assigned to treatments (i.e.

institutions) randomly, i.e. the researcher could not randomize which variables (institutions) were assigned to which unit. The variables that are held constant for this study are therefore not "controlled for" in the strict, experimental sense; however, the two compared cases were selected to be as similar as possible to make the claim of constant values of variables that are not tested for. In this respect, here we can speak of a natural experiment, implying that "nature" drew the borders between the two states on Borneo and the same society, in particular the same ethnic composition, was "treated" with different institutions, enabling us to observe how the same ethnic composition varies in terms of activated identities under different institutional designs.

An attempt to answer the questions listed earlier requires one of two possible approaches: either a study of a political unit over time, with changes of the institutions between electoral periods and data reflecting the change of identity, or a comparison of political units with different institutions but similar ethnic categories. Daniel Posner (2005) and Kanchan Chandra (2004) conducted their research according to the first approach, with their main focus on parties. Sub-national units of Zambia and India, respectively, were analyzed over time and conclusions were drawn from within-case analyses. The current work, on the other hand, is an attempt to frame the analysis not only on the temporal analysis (change over time), but also on the second approach: sub-national units of Malaysia and Indonesia will be compared to establish the impact of institutions, power structures and political parties on ethnic identity change.

Hence, this study is conceived as a comparative one, and corresponds to type 2 of the comparative studies identified by Peters, which encompasses "analyses of similar processes and institutions in [a] limited number of countries, selected (one expects) for analytical reasons" (2004, 8). The present one is a comparison of two sub-national level units that is hinged on institutions as independent variables. Przeworski and Teune point out the perennial dilemma between studies that are system-specific accurate and studies that are less accurate, but offer potential for generalizations (1970, 22–23). This work is tipped towards the accuracy pole, as it immerses deeply in the historical and situational specificity of the studied cases. Generality of the results presented here is clearly limited, but, in accordance with Mahoney, there is an upside to it: although "the constraint on generality in small-N analysis is a significant limitation, it can often be balanced by the substantial conceptual gains derived from in depth knowledge of particular cases" (2000, 86).[12]

Malaysian and Indonesian ethnic mobilization has a history that stretches the entire period of the countries' existence. In fact, in the case of Malaysia, the ethnic negotiations *are* the history of political dealings in the state. Throughout the entire process, the formal institutions changed once: when Sarawak joined Malaysia and adopted its electoral system. Beyond this point, the formal conditions of the game were kept constant, but two later events marked the establishment of informal, yet rigidly observed new rules. In Indonesia the institutions have changed frequently, in either an evolutionary (during the New Order and later since 1999) or a revolutionary (1965, 1998) manner. As will be shown, however, the impact of these changes on identity change in West Kalimantan until 1999 was minimal.[13] Therefore, ethnic identity activation is a phenomenon that lends itself well to a historical-institutional analysis. In both Malaysia and Indonesia, we can speak of perpetual *processes* that can be *traced*, of *paths* on which new developments *depend* and of *junctures* that are *critical* to ethnic developments in politics that can be observed. The time and sequence of the events are all important here. Therefore, the analysis of both cases follows historical institutionalism, which looks out for "organizational and institutional configurations [. . .], critical junctures and long term processes" (Pierson and Skocpol 2002, 693). These in turn become what came to be known as path dependence, or "the dynamics of self-reinforcing or positive feedback processes in a political system. [. . .] A clear logic is involved in such path-dependent processes: outcomes at a critical juncture trigger feedback mechanisms that reinforce the recurrence of a particular pattern into the future" (Pierson and Skocpol 2002, 699).

In the case of Sarawak, it will be observed that the early events, especially the removal of the first chief minister from office in 1966 and the creation of the first coalition government in 1970, made the actors "venture far down a particular path, however, they [. . .] find it very difficult to reverse course" (Pierson and Skocpol 2002, 699–700). Ethnic dynamics in Sarawak beyond this juncture came to exclude several hitherto possible options, or as Pierson and Skocpol put it, "political alternatives that were once quite plausible may become irretrievably lost" (2002, 700). As a result, particular political configurations became institutionalized. Therefore, there is a lot of emphasis on early ethnic identity changes in Sarawak, and newer events are studied chiefly to demonstrate their consistence with or deviation from the established pattern. The case of West Kalimantan is balanced in a different way. Very little change was observed in the first 50 years after independence. Categories activated during the Old and New Orders hardly changed, and the political environment and institutions produced a highly consolidated, single-dimensional set of identities with which the West Kalimantanese identified. The search for identity shifts is therefore focussed on the most recent elections, taking Suharto's fall as the critical juncture, which is expected to change the previously existing pattern.

Although process tracing is also applied to the Indonesian case, in itself this approach would not yield high results. As I expect to find most of the identity shifts taking place during the past two electoral cycles, the focus in Indonesia is not on the timeline as much as in Sarawak. The reason is quite simple: there were only short periods of free political mobilization and competition in Indonesia, and almost 40 years (the last years of Sukarno and the entire period of Suharto's regime) cannot be analyzed in terms of free political activities. Party-based mobilization took place for a short while and presented during two elections (general and provincial) in the 1950s. But later the chronology developed outside of electoral institutions or parties; ethnic mobilization during this period took place in the form of repeated ethnic violence (Davidson 2008a) and a path of ethnic activation was then established. Ethnic categories in West Kalimantan were activated and fixed within violent episodes. As this trajectory was established by

Davidson (2008a), it will be merely presented here as point of departure for the later developments.

The quick and dramatic change of institutions after 1998, however, relaxing of the state propaganda and the regime's tight grip, as well as the blossoming emergence of new administrative units with potential new ethnic distributions and loyalty lines, was the critical juncture that could change the trajectory of the path. The post-Suharto period called *Reformasi* (Reformation) provides therefore an opportunity for a different focus of analysis. With this historic turn, we should assume new patterns of development of ethnic mobilization. Specifically, knowing the set of activated categories towards the end of Suharto's rule, the crucial problem to study is: were the new institutions and new opportunities for mobilization able to break out of the mobilization pattern from the New Order? Does ethnic identity activation continue to move along the same path established by violence? Or could new parties and new potentials for minimum winning coalitions lead to new patterns of ethnic identity activation and frequent shifts between activated categories? To answer these questions, which pertain to events of a mere 15 years, we will be looking at elections on each of the administrative levels to study ethnic change between districts, between a district and the province, a district and the national level and so on. It will be a much more spatial study of elections in the case of Indonesia, but focussed on the past two election cycles, unlike the case of Malaysia – 50 years and 10 election cycles.

1.3 Research questions, hypotheses and more theoretical considerations

Earlier in this chapter I proposed that the two cases studied here be classified according to each state's approach to the ethnic divisions in its society. The general distinction between the two cases, I argued, could be framed as consociationalism versus centripetalism. In Malaysia the ethnic divisions are recognized, politically accommodated and enhanced within a consociational polity. In Indonesia, state institutions are designed to minimize the opportunities to mobilize any ethnic categories except for the religious ones, which, combined with state ideology, makes Indonesia a centripetalist design. This binary distinction between cases, while it neatly frames the design and helps generate hypotheses, does not exhaust the explanatory potential embedded in the institutional systems of the two cases. If we take apart the institutional structure and look into each particular element of it, we will be able to pin down not only the causal effects of institutions as incentives for ethnic identity change, but also we will escape the trap of an over-simplifying binary distinction. Taking advantage of the descriptive nature of a small-N study, the finer points of these two institutional settings will be traced and interactions between each of these institutions will be exposed to unveil causalities that may not be seen in a static observation. The bottom line is that the independent variables are expected to be interwoven, and causal inference will be flawed by the difficulty of treating these variables as acting independently from each other. In accordance with this, questions asked in this research and hypotheses refer more to the combined effect of the variables presented later; causality will not only be attributed to the individual variable, but also to the synergic end result of multiple elements of the entire institutional setting.

Principally, institutions are not exogenously given[14]; on the contrary, political entrepreneurs have enough incentives to manipulate them. Indeed, as Indrayana (2008) and Horowitz (2013) showed using the Indonesian example, political entrepreneurs strongly influence the shape and structure of institutions. However, whatever bargaining happens at the centre, it is not influenced by the provincial/regional/local power relations or ethnic entrepreneurship. A study on the sub-national level of politics and a choice of quite remote, parochial provinces legitimizes the otherwise troublesome claim of independence of institutions as variables. The structure, type and relations between political institutions were merely passed down from the centre onto provinces of Indonesia and states of the Federation of Malaysia. Sarawak is a particularly clear-cut case: the nature of the state and the structure of political power were all established before Sarawak became part of Malaysia.

While I acknowledge that the colonial administration shaped the two regions in some way, for practical reasons I do not focus on the influence of the pre-independence administration; the emphasis here is on post-independence developments. Nevertheless, electoral systems of the two states were clearly inherited from their respective colonial powers and so Malaysia maintains single-member constituencies with plurality vote, with indirect executive elections, while Indonesia upholds multi-member constituencies and proportional elections, with the executive heads elected directly on three tiers. We take 1963 (for Sarawak) and 1945 (for West Kalimantan) as $t = 0$ of institutional developments and the study maintains sensitivity to institutional alterations[15] and their influence on identity category activation. Indisputably, the first ethnic category activation occurred prior to independence and these categories will be taken into account as the point of departure.

Because we take the constructivist paradigm seriously and follow through with its assumptions, we will ask questions that position ethnic identity as the dependent variable, and – given that we are operating in the field of social sciences – ask how *institutions* drive ethnic identity change. Table 1.1 presents the institutional setting of the two cases. This dry matrix serves the purpose of showing how differently the institutions were designed, but explains little about practical ways of political proceedings in the two cases. In both countries the institutional setting has to be shown as a process of evolution, and not as a static given. Therefore, each of the empirical chapters will track ethnic identity change parallel to institutional changes to demonstrate their dynamic nature and the developments in political practice.

Equipped with the powerful assumption that one can choose between multiple identity categories, e.g. language, race, religion, tribe, region, and having assumed that this choice can be influenced by incentives, we ask: what are the incentives that lead to ethnic identity change? This study will attempt to enrich our knowledge of ethnically relevant incentives by focussing on political institutions and rules for political competition. The analysis will revolve around questions like:

Table 1.1 Political institutions of Malaysia and Indonesia

	Malaysia	Indonesia
Structure	Federal	Unitary decentralized
Political system	Parliamentary	Presidential
Legislative	FPTP, two levels	PR, three levels
	State assembly and national parliament directly elected; members of the Senate nominated by the Yang di-Pertuan Agong and the state legislatives	District-level assembly, province-level assembly, bicameral parliament (both houses directly elected)
Constituencies	Single-member; re-delineated by the Election Commission every eight years; gerrymandering relatively easy as the shape of constituencies is not tied to other units	Multi-member; drawn along administrative units, i.e. gerrymandering only possible via creation of new administrative units
Executive	Indirectly elected, ministers on both levels are all members of the respective legislative (including members of the Senate). Two levels	Executive head directly elected at three levels on dual ticket (officeholder and deputy), to win 30% votes in first round needed or run-off; candidates need to be supported by party or party coalition that obtained at least 15% of votes or seats in the last election on the relevant tier. Non-combination of legislative and executive positions.
Political parties: regulation	No limits in respect to ethnic mobilization; operation of federal-level parties subject to registration in respective states	Ban on regional parties (except Aceh), only nationwide parties compete in elections on all levels; religious parties free to operate

Source: Author's compilation.

What are the mechanisms of identity change under differing institutional structures? What elements of these structures matter the most? What institutional arrangements speed up activation of categories, and which institutions can induce more frequent shifts between categories within an individual's repertoire? Finally, the most exciting question seems to be: is the change of ethnic identity faster in plural societies with more elected offices and varying electoral districts for each election, and are the shifts of ethnic identity more frequent in such societies? Do consociational systems, by contrast, indeed arrest ethnic identities by making them explicit and inducing perpetual and constant activation of the same categories?

Let us point out an important element of the ethno-political reality: ethnic identity may be "fixed" or "fluid", but not because of its intrinsic quality of ethnic identity but because of the institutional setting. Within near-homogeneous, single-member districts we may witness a process of *institutionalized fixedness*, where only one category is activated over dozens of years. If, however, we observe multiple elections on different administrative levels and for different offices, we expect to see shifts of identities among voters from one election to another, depending on the district magnitude and the position that is filled via the election and party strategies.

Earlier I presented the designs for divided societies and how the cases analyzed here fit into those designs. Based on them, we can identify our expectations on ethnic identity change in polities with these institutional designs. Kanchan Chandra (2008) proposed that societies in which each person has in her ethnic repertoire more than one activated category and has incentives to switch between these categories can be more peaceful and stable democracies. Chandra argued that the more elections take place on different administrative levels, the more opportunities exist for the activation of alternative categories and shifts between them. We will use Chandra's proposition as a ready hypothesis for this research, having at our disposal one case with few elected offices, and another case with as many as seven elected offices.

H 1a Fewer directly electable offices result in the activation of fewer categories in individuals' repertoire or less frequent shifts between them.

H 1b More direct elections induce a higher number of activated categories in individuals' repertoire across the society and more frequent shifts between them.

The second hypothesis corresponds to our expectations of speed of ethnic identity change depending on the formally or informally permitted forms of mobilizing ethnic categories. Some polities – most prominently, consociational ones like Malaysia – thrive precisely in the context of explicitly spelled out ethnic categories. If ethnic mobilization is frowned upon and excluded from public presence – think of paradigms behind centripetalist institutions, and the case of Indonesia – implicit mobilization is the only way for ethnically motivated political entrepreneurs. An implicit message, however, requires more time and sophistication to be conveyed, and is less precise in its contents, while an explicit message is quick, straightforward and precise about its contents. Therefore:

H 2a Explicit mobilization of ethnic categories results in higher speed of identity change.

H 2b Implicit mobilization of ethnic categories results in slower speed of identity change.

The third hypothesis draws again from differences between consociationalism and centripetalism. In the case of consociational designs, it is assumed that it is through the elite bargain that power is shared, and electoral outcomes cannot change the conditions of the bargain. This is expected to be true for several electoral cycles: an ethnic bargain is established to serve over decades, and not just for the current election. Centripetalist designs include no elite bargains and leave the ethnic composition of the legislative and executive up to the decision of voters in each election. As much as centripetalism tries to render ethnicity an irrelevant factor in politics, Indonesia and other countries clearly demonstrate that this design often fails to keep ethnicity out of the political equation. Therefore, "failed centripetalism", where ethnic competition is not eliminated despite institutions designed to do so, can be a hotbed for ethnic identity change, as each electoral cycle represents a new occasion for mobilization. Therefore:

H 3a   A consociational polity with power-sharing bargains between ethnic elites arrests ethnic identities by perpetuating activation of the same categories and dimensions over time and elections.

H 3b   Electoral competition devoid of power-sharing schemes induces speedy activation of different categories and enables shifts between them.

It is important to notice that among the cases studied in this research we have one centripetalist with a high number of elections and implicit mobilization (West Kalimantan in Indonesia), and one consociational with few elections and explicit mobilization (Sarawak in Malaysia). According to hypothesis H1, Indonesia would have faster and more frequent identity change and more activated categories in every person's repertoire, but according to H2, Indonesia would see slower identity change due to implicit ethnic mobilization in this country, and according to H3, the opportunities of activating alternative categories from each person's repertoire should be more in Indonesia. Malaysia is the exact opposite: the few elected offices are expected to pose the country to have fewer opportunities to activate new categories, but thanks to the permitted explicit mobilization the chances of ethnic identity shifts should be higher. However, the general expectation of consociationalism is that it is based on one set of categories perpetuated over time, making us expect Malaysia to have no more than one activated category for each individual. Therefore, each of the institutional elements may lead to a different outcome.

In a qualitative study with only two cases, such as the present one, the gain lies in tracking precise mechanisms that drive the ethnic changes, and not in producing universal truths about them. This study involves various types of data and is immersed in contexts; therefore there is a chance here to take into account facts as they present themselves without sifting them out a priori. Consequently, parallel to the theory-driven hypotheses presented earlier and analysis that tests them, rich empirical particularities will be shown to interfere with the expected outcomes. In particular, political parties and federal–state relations in the two cases will undergo new scrutiny and their role in ethnic identity change will be shown in a new light.

This may not only enrich the explanatory power of this study, but also contribute to the existing conceptual framework. We may be able to learn from Sarawak and West Kalimantan about ethnic strategies of parties in consociational settings, as well as, drawing from Indonesia, ethnic strategies of parties under ethnic party bans,[16] which can help to refine the concept of ethnic parties. In a similar way, I expect to watch the different ways the federal/central government interferes (or refrains therefrom) in regional ethnic negotiations. A commonsensical expectation that a federal government would leave a free hand to ethnic bargaining in individual states will be probed in the case of Malaysia. The alternative setting – a unitary, albeit decentralized Indonesian state, will offer a convenient case to test the central government's behaviour in the case of lack of autonomous provisions in particular provinces.

These two factors and their impact will be studied in an exploratory way; it is not hypothesized about them prior to the analysis. The rationale here is that the role of these two variables in the respective cases, and the combined impact of these variables, may go beyond what we may assume based on the current knowledge. An exploratory approach will allow a study of all the available facts and avoid being misguided by too narrowly drawn hypotheses. Hopefully, new insightful findings pertaining to the impact of parties and federal–state relations will present themselves upon completion of this study.

Therefore, within each case we expected to find enough reliable evidence to confirm or deny the hypotheses outlined earlier, or otherwise find additional factors that interfere with the assumed causalities. It is, however, necessary to keep in mind that the explanatory power is limited to these two cases. In this respect we are dealing here with an exploratory research that seeks accuracy of analysis and uncovers mechanisms behind ethnic fixedness and fluidity that are more sophisticated than what can be expected of a mechanical analysis of dichotomous variables. Moreover, answering the questions listed in previous paragraphs and testing the identified hypotheses will help make informed predictions about the cases analyzed in this study and will bring us closer to concluding about preferred institutional design for plural societies.

The concept of ethnicity and ethnic categories used throughout this work is drawn from the extensive work of Kanchan Chandra, who not only proposed a new theory of ethnic identity change but also sharpened the edges of the conceptual framework of ethnicity:

> [E]thnic categories are a *subset of categories in which descent-based attributes are necessary for membership* [. . .] *all* ethnic identities require *some* descent-based attributes for membership. *Nominal* ethnic identities are those ethnic identity categories in which an individual is eligible for membership based on the attributes she possesses. *Activated* ethnic identities are those ethnic categories in which she professes membership, or to which she is assigned by others as a member.
>
> (Chandra 2012c, original emphasis)

Chandra further explains that attributes associated with descent are those that are either genetically inherited (e.g. skin colour, hair type and other physical features), or through cultural inheritance (e.g. names, languages, place of birth and origin), or through cultural markers acquired during one's lifetime (e.g. last name or tribal markings). The rules of membership in identity categories, elaborates Chandra, are either explicit or implicit and do not need to be uniform. Moreover, "the existence of explicit and uniform membership rules [. . .] is not a random event: it is correlated with political and economic circumstances which attach rewards or punishment to the codification of these rules" (Chandra 2012a, 13 fn).

These specificities are particularly important for Malaysian Muslims; adherence to Islam and habitual usage of the Malay language are constitutionally defined as sufficient to be classified as Malay in the Malaysian constitution. The situation in Indonesia is less rigid and potentially more confusing. Converting to Islam used to be a sufficient criterion to enter the Malay category. In modern Kalimantanese politics, conversion does not automatically and unconditionally suffice to obtain membership in the Malay category. What is, however, more important is the understanding in the two countries that the religious dimension in Borneo is not considered based on descent alone and there is an element of choice. By the principles of the definition of ethnic identity given earlier, converts would not obtain membership in descent-based categories, as their religious affiliation is not based on descent. Nevertheless, I allow the particular membership rule (which has widespread acceptance) to precede the definitional constraint; this is also in line with the political practice in Malaysia and Indonesia.

The change of ethnic identity is constrained, continues Chandra, by the inherited attributes: dyeing of one's hair, learning of a new language or moving to a different region does not suffice to claim membership in new categories. Therefore, the change of identities is constrained to the ones that are available through membership rules, which in turn are constituted by a fixed set of inherited attributes. All categories in which one can claim membership constitute one's ethnic repertoire and within this repertoire an individual can shift identities. Another property of ethnic identities Chandra identified is visibility, which for her means that "some information by which an observer can place an individual in an ethnic category is available through superficial information, although there may well be variability and error in how this information is interpreted" (2012a, 2).

In sum, an individual changes her identity by picking one category from the set of categories in which she is eligible for membership based on fixed attributes that are unchangeable in the short term. If we assume that the shift between the categories happens, among others, because of "the short term incentives imposed by political institutions, then our theories of political competition cannot take ethnic identities as fixed and exogenous even in the short term" (Chandra 2012c, 8). This study precisely follows this argument and assumes changes of identity happening constantly and within short periods of time.

Political parties are, on one hand, an agent of mobilization and consequently of identity change; on the other, they are a product of the existing political institutions. The focus on parties in this study will revolve around the question whether the parties in Sarawak and West Kalimantan are ethnic, and how they impact ethnic identity change. To establish whether parties are ethnic, parties will be checked for three properties, as Chandra proposed: particularity, centrality and temporality of interests championed by the party (2011, 155). By "particularity" it is meant "that an ethnic party as defined here must always exclude some group, implicitly or explicitly" (Chandra 2011, 155). Centrality refers to a situation when "interests of some ethnic groups is central to the signals a party sends, [and therefore] this definition rules out parties that make only peripheral references to ethnic categories" (Chandra 2011, 155). Temporality allows for the party to change its ethnic outlook over time: "The group or groups that a party speaks for can change across time. Consequently, whether or not a party is 'ethnic' in nature can also change over time" (Chandra 2011, 155). Spatial analysis of parties in Indonesia may add to Chandra's definition of parties. In this work it will be checked to what extent parties can be locally ethnic, or better, whether parties that are otherwise non-ethnic can have ethnic party properties in particular areas or on certain electoral tiers. If it is indeed found that some parties fulfil all criteria of an ethnic party but only within a province or regency, but do not fulfil such criteria on the national level, the question should be asked of whether the requirement of exclusion – difficult to reconcile with the "locally ethnic" property – better be relaxed to make sure that the definition does not eliminate parties that display ethnic properties in regions, but not on the central level. Indonesia's PDI-P (Democratic Party of Indonesia – Struggle) is one example of such party and one that can significantly add to our understanding of how parties operate.

Furthermore, argues Chandra, there are eight indicators according to which an ethnic party can be recognized: 1) its name, 2) categories explicitly advocated for in the party's campaign message, 3) issues explicitly advocated for in the message, 4) the party's implicit campaign message, 5) the groups who vote for the party, 6) the composition of votes the party obtains, 7) composition of leadership and 8) its arena of contestation (2011, 157). Analysis of party politics will follow these indicators and, based on them, the nature of parties in the two cases studied will be assessed.

This work draws from two contributions by William Riker. Foremost, his study of coalitions as formal processes and for formalized purposes (e.g. government creation, establishing of a majority in an assembly to pass a bill) is invaluable for the current analysis. The other one is his study of sophisticated and rather non-formalized, elusive political strategies; what he called heresthetics is a technique of manipulating of one's preferred outcome, without changing underlying preferences. The technique is clearly vested in rational-choice theory. Riker starts with defining a "politically rational man", who in Riker's words is "the man who would rather win than lose, regardless of the particular stakes. [. . .] The man who wants to win also wants to make other people do things they would not otherwise do, he wants to exploit each situation to his advantage, and he wants to succeed in a given situation" (Riker 1962, 22). Consequently, the ethno-political entrepreneurs studied in this book will be considered politically rational people who deploy ethnic identities to arrive at preferred outcomes. Manipulation of preferences is a useful tool in the hands of a politically rational man. As Riker explains the process, "for a person who expects to lose on some decision, the fundamental heresthetical device is to divide the majority with a new alternative, one that he prefers to the alternative previously expected to win. If successful, this maneuver produces a new majority coalition composed of the old minority and the portion of the old majority that liked the new alternative better" (1986, 1). This simple rule will be shown to guide mobilization strategies of political parties and political entrepreneurs who seek to create new ethnic minimum-winning coalitions, or, in terms of the theory presented earlier, try to activate such ethnic categories that reshuffle previously winning ethnic categories, so that the new categories are more preferable than the previous ones.

Riker's main work on coalitions, "The Theory of Political Coalitions" (1962), deploys game theory[17] and the size principle to explain mechanisms of coalition building. In terms of game theory, coalition building is an *n*-person game, or simply there must be three or more players so that any further analysis of coalitions is of relevance (Riker 1962, 35–36). Riker notes that "evidently there are some restraints operating on such persons so that the actual choice among coalitions is limited" (1962, 36); in real-life situations, the restraints correspond to e.g. exclusion of communist parties from coalition negotiations in some countries, or particular restraints to exclude or mandatorily include a particular ethnic category in a government coalition. Engineering ethnic vote always involves establishing coalitions, both in a formal way (parties agreeing to nominate a common candidate for an executive post or to enter a government coalition), and in an informal way, by deciding which ethnic categories of the population to target.

Coalitions are categorized as follows:

> [A] *winning coalition* [. . .] is as large as or larger than some arbitrarily stated in the rules. All coalitions that are not winning are either *blocking* or *losing*. The complement of a winning coalition is a losing one. The complement of a blocking coalition is a blocking coalition. A *minimum winning coalition* is one which is rendered blocking or losing by the subtraction of any member.
>
> (Riker 1962, 40, original emphasis)

This categorization will inform all types of coalitions in this work: whether coalitions between parties or ethnic categories (e.g. in Indonesian executive elections, expressed in the ethnic categories of contesting pairs of candidates); all will be winning, minimum winning, losing or blocking. This book will present relatively few minimum-winning coalitions – most of those presented here, regardless of administrative level, will be highly inclusive coalitions that in terms of both parties and ethnic categories represent a larger number of members than is necessary to maintain power. We will provide some answers as to why ethnically defined coalitions might be prone to being oversized, and how it might have a bearing on political stability in polities which experience this phenomenon.

The electoral process is additionally ruled by the principles of strategic coordination, which stipulate that "(a) candidates wish to get elected and voters wish to gain the benefits of voting for winners, and (b) candidates' and voters' expectations of winning and losing tend to be mutually reinforcing" (Mozaffar, Scarritt and Galaich 2003, 380). This has immense consequences for our understanding of the ethnic vote. While it is generally assumed that people prefer to vote for their co-ethnics, they should also prefer to choose the winning candidate, and he might not belong to any of the voters' ethnic categories. This book will show several examples of support for winning candidates *against* the ethnic rule. At the same time, I will show that an ethnic category's interests within a given constituency can very well be secured by a representative who shares no ethnic membership with the voters, and yet, because he wants to be (re)-elected, he will be as politically useful to his constituencies as an ethnic representative would. The same caveat will be observed for parties. Despite their ethnic commitments, occasionally they will be shown to back a candidate from outside of their ethnic bases, if that candidate is the most likely winner.

Cox's research informs that "electoral coordination occurs at two main levels: (*a*) within individual electoral districts, where competitors coordinate entry and citizens coordinate votes; and (*b*) across districts, as competitors from different districts ally to form regional or national parties" (1999, 145). These observations make us look at parties, candidate nominations and post-electoral negotiations through at least two and often three different lenses: on the local/constituency level, on the level of the regional government (e.g. state in Malaysia and province in Indonesia), and on the national tier. Undoubtedly, each election has a bearing on the power constellation on each of these levels and coordination of preferences of political players on all these levels can lead to quite interesting solutions. Opposition parties on the state level will be shown to be retained in the federal cabinet; in other instances, parties cooperating on the national level refuse to enter a coalition on the provincial level because of their ethnically defined constituents. To add to this complexity, I will argue that a party can have a clear ethnic outlook at one level of the political tussle (a province), and be a non-ethnic party elsewhere, or represent a different ethnic category.

To sum up, this study tries to collect convincing evidence of ethnic identity change and, more specifically, how institutions impact the change. I approach the problem by studying two neighbouring provinces of two states; the provinces (but not the states) conveniently have very similar ethnic structures (set of nominal ethnic categories) and very different political systems that serve as the independent variable. Therefore, I expect to see

1   Different identity categories being activated in both provinces (i.e. differing ethnic practice) because of the different institutional design of the political realm.
2   Varying speed and frequency of identity change.

The results of this study point at interesting phenomena. Sarawak turned out to be a polity in which each individual retains at least two or three activated categories in her repertoire, and Sarawakians are frequently induced to shift between these categories within short periods of time. In Sarawak, the elite bargaining happens through at least three channels (political parties, legislative seat assignment and executive nominations) and through each of these channels different ethnic categories can be activated. This results in the frequent ethnic identity change. Individuals shift between categories from a dimension that combines religion and origin ("Muslim indigenous", "non-Muslim indigenous" and "Chinese"), dimension of region ("Bidayuh", "Iban" and "Orang Ulu") and dimension of language ("Malay", "Melanau" and several categories within "Bidayuh" and "Orang Ulu"). Shifts between these categories, however, are strictly conditional on explicit ethnic mobilization, and are a result of particular historical development of institutions. Political parties in Sarawak are found to be a channel of activation of several categories that very likely would not have been politically mobilized otherwise. Therefore, consociational polity cannot be automatically associated with arrested ethnic identities.

In Indonesia, ethnic identity change was found slower and less frequent than expected. Liberalization of political life after 1998 did not lead to widespread activation of new categories in people's repertoires. On most occasions voters in West Kalimantan still identify only with categories activated during the Sukarno regime. Political parties do attempt to and succeed in circumventing the regional party ban, but they also seek to mobilize the long-entrenched categories. The direct elections at the sub-regional level in some districts induced activation of hitherto only nominal categories (i.e. linguistic, or Christian denominational, like Protestant and Catholic), and in these districts a second category was activated in individuals' repertoires. Significant potential of ethnic identity change in West Kalimantan lies in these district-level elections, but in many of them there is strong ethnic inertia, focussing the ethnic politics around the Muslim versus non-Muslim division.

1.4 Controlled variables: democracy, patronage and clientelism

The two cases studied here represent the "most similar systems design", and inter-systemic similarities and inter-systemic differences are the focus of this study. "Systems constitute the original level of analysis, and within system variations are explained in terms of systemic factors. [. . .] Common systemic characteristics are conceived of as 'controlled for', whereas intersystemic differences are viewed as explanatory variables" (Przeworski and Teune 1970, 33). This study is primarily concerned with sub-national units of states. This has important consequences for how we see the variables. As mentioned earlier, dealing with sub-national units allowed me to consider institutions as exogenously given, which on the national level could be troublesome. Later I present controlled variables that I hold stable for the two cases for the sake of this research, but have to admit that even within each of the two states, and particularly between the national level and the sub-national level, these values do not have to be stable. Depending on region, levels of urbanization, education and penetration by the media and other factors, we may find great variance in levels of political freedoms and dependence on patronage politics within one state. Moreover, the contents of political debates should be expected to change between the national level and sub-national level.

The uneven levels of democracy in the two polities pose a serious methodological and epistemological challenge to this research. Malaysia has held competitive elections across the entire period studied here, but since 1970, none of these elections has been free or fair.[18] Malaysia consequently falls into the "partly free" category in Freedom House's classification. Indonesia did not hold democratic elections between 1959 and 1998 and this entire period will only be discussed here as background of some phenomena and root cause of others, but I provide no new interpretation of that period, as no meaningful identity change due to institutional conditions happened during that period, or none can be traced via election analysis. Since 1999, however, Indonesia has had several free, fair and democratic elections and since 2005 Indonesia is a "free" polity according to Freedom House (compare Figure 1.1). What do these discrepancies mean for a comparative study? Several caveats have to be made before we can legitimately draw conclusions from electoral results in "democracies with adjectives".

The Polity IV data set (Figure 1.2) may help to make the point here. On this scale democracy and undemocracy (Tilly 2008) take positive and negative values, respectively. Polities scoring below 0 are non-democratic and no meaningful conclusions on voters' preference can be drawn from electoral results during these periods (even if elections are held, *vide* the case of Indonesia under Suharto). Even when we eliminate these periods from the study, we are still dealing with very different levels of democratic freedom in the two cases. To use Przeworski's (2007) pregnancy metaphor, if a country's democracy is between -10 and 0, it is much like a "not pregnant woman". Between 1 and 10, however, a country is

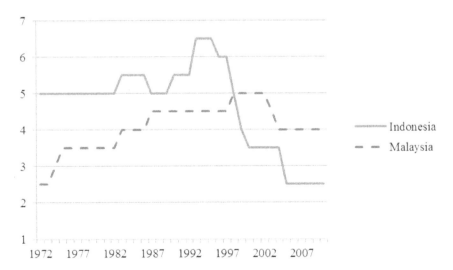

Figure 1.1 Political freedoms in Malaysia and Indonesia according to Freedom House

Source: Author's compilation based on FreedomHouse.org. Countries' "political rights" and "civil liberties" are rated on a scale from 1 to 7; the final score is the average of the two values. 1 denotes "free"; 7 denotes "non-free".

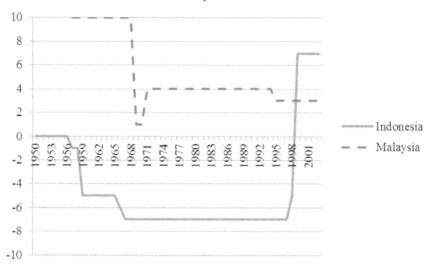

Figure 1.2 Political freedoms in Malaysia and Indonesia according to Polity IV Project

Source: Author's compilation based on www.systemicpeace.org/polity/polity4.htm. The "Polity Score" captures regime authority spectrum on a 21-point scale ranging from -10 (hereditary monarchy) to +10 (consolidated democracy).

democratic, but to a differing extent, much like a woman can be further along in her pregnancy. By this rule, Malaysia has been in the first trimester of its democratic pregnancy for more than 40 years; Indonesia was not pregnant throughout the entire Suharto rule (1965–1998), but jumped to the third trimester within a few years after that. Indonesia has succeeded in its transformation, but has not come to full term yet.

An important light may be shed on the situation by asking what the major flaw of the democratic process in the two states is. The list of flaws in the electoral process in Malaysia is long, but my informed estimate based on extensive study of the country suggests that the most effective measure of inducing the desired outcome of elections in Malaysia is patronage. Simply put, money trumps all other manipulation tools available. Interestingly, if we turn our scrutinizing eye onto Indonesian elections, we will find that, here too, it is the financial incentives that have the most convincing power (Hadiz 2010; Subianto 2009).

Consequently, elections on both sides of the Indonesian-Malaysian border have to be discounted by the value of the patronage factor, and on the sub-national level equally as on the national level. Institutionalization of patronage politics may seem a strong argument weighing for greater significance of the said factor in Malaysia. On the other hand, as Case (2011) points out, the Malaysian legislature, with its fearless and outspoken (albeit less so in Sarawak than on the national level) opposition, has much greater controlling power than Indonesian legislatures on all levels. Cartelization (Katz and Mair 2009) of Indonesian political parties renders "opposition" an almost meaningless concept in Indonesian reality (Slater

2004), and there is little incentive for any party to challenge the patronage-driven mobilization. Mietzner challenges the cartelization claims on the grounds of clear ideological differences between parties (2013, chap. 5). Mietzner shows that parties in Indonesia differ significantly, especially on the spectrum of Islam versus secularism, and voters' preferences reflect this dimension of party ideologies very well. Chapter 5 of this book presents further evidence to support both claims: parties do form a cartel in the sense that all parties participate in the spoils of power access at all times and there is little accountability resulting from opposition parties keeping tabs on the ruling coalition.

At the same time, with clear ethnic preferences reflected in party support, parties' ideological stances are not irrelevant and continue to shape voters' attitudes. Moreover, in the case of West Kalimantan, the ethnically derived division between Golkar and the PDI-P resulted in the two parties consistently competing against each other and never forming coalitions in the province and regencies, although they almost indiscriminately accepted all other parties in their coalitions at different times. Therefore, ethnic parties in ethnically competitive polities are logically non-cartel. At the same time, once elections are over, those same parties show no opposition-like behaviour, which implies their cartel nature. Having acknowledged these important nuances related to party behaviour, for the sake of carrying out this research, I have no choice but to assume that both countries are value 1 on the binary scale of patronage (0, 1), with 1 corresponding to patrimonial political relations and 0 to patronage-free political life.

While recognizing the long-proven impact of (neo-)patrimonial relations on ethnicity (Bates 1974; Scott 1972a, 1972b), patronage is not an exogenous variable in this research. The rationale is as follows: for one, the aim here is to test the formal institutions and their role in setting the conditions for identity change. For two, there are excellent studies focussing on ethno-patrimonialism: Chandra (2004) finds patronage to be a strong explanatory variable in ethnic party mobilization, while Fearon (1999) shows why patronage and ethnicity go hand in hand in electoral mobilization. Although Malaysia and Indonesia have great potential to be a laboratory for ethno-patrimonial studies, the long overdue focus on formal institutions is much more urgent.

This work, however, makes an assumption about the role of patronage in the ethnic bargain: patronage is expected to trump any ethnic loyalty (which in an ideal case would make ethnic identity endlessly fluid, or politically irrelevant), but it would be financially unsound for political entrepreneurs to continue to buy loyalties.[19] Money, as a limited resource, is used only when absolutely necessary. Cases of monetary gains overruling ethnicity exist, but remain an exception; as a rule, patronage is an accompanying means to tip the balance towards the desired outcome, not to turn it upside down. To conclude, despite the differences in democracy levels, we need to acknowledge flawed electoral processes in both cases. Yet, as long as elections remain at least minimally competitive, i.e. a voter can express at least some preference, the election is of significance for this research.

Having mentioned the preference of a voter, we also need to ask another important question: why is it assumed that electoral results in the two countries reflect voters' ethnic affiliation? Why is it expected that ethnicity plays any role at all in an election? Is it not primordial to assume that ethnicity *must* be not only *a* cleavage, but also *the* cleavage? Although I do pose the question of which of many ethnic identities is activated in any given election, at the same time I seem to assume that other cleavages are not relevant enough in these polities to overshadow the ethnic outlook of elections. This point has to be tackled here, and from more than one perspective.

One perspective originates from Chandra's (2004) "limited information" theory; according to her, information about the ethnic background of a candidate is the only information that a voter can obtain without any cost. Language, clothing, name, physical appearance all tell us plenty about a candidate's ethnic background, but reveal no (or little) information about a candidate's class, education, profession or ideological stance. In this light, in communities and societies where obtaining any information in general is relatively costly (because of poor infrastructure and consequently poor transport and communication, lack of independent media or any media, illiteracy etc.), of which Sarawak and West Kalimantan are good examples, mobilizing along the very visible descent-based affiliations is a likely outcome. Simply put, in predominantly rural regions voters usually know not much more about their candidates than their ethnic background. Because the candidates are well aware of this information gap, arguably they are also prone to take advantage of the free-of-charge mobilization tool. Therefore, there is nothing primordial in assuming the importance of ethnic affiliation in electoral choices; on the contrary, it is an assumption based on rational-choice calculations.

A similar argument was developed by Mozaffar and colleagues, who observed that

> In emerging democracies, however, electoral institutions are new and their incentives and outcomes not well known or understood by political actors, who compensate for the resulting information deficit by relying on alternative sources of information and coordination. In Africa, ethnopolitical groups and cleavages are these alternative sources.
>
> (2003, 380)

The early years of Malaysian party dynamics and the most recent developments in Indonesia suggest that Mozaffar's conclusions in Africa correspond to the situation in the two Southeast Asian countries.

The second perspective is a cleavage model–based argument; the cleavage model (Lipset and Rokkan 1967) would suggest class and urban-rural divisions as potential mobilization sources in the said societies as an alternative to descent-based cleavages. We would be asking, is either of the two cleavages to be found in the two polities? A brief overview of policy propositions brandished by parties in Sarawak and West Kalimantan, and a quick check of political debates over the years in the two states, convinces us to discard these two cleavages, even if some of the parties do mobilize according to these cleavages at the national level (compare Ufen 2008a). Sarawak is a false positive when it comes to the relevance of the rural-urban division. Here, the cities and countryside display very strong discrepancies in their voting patterns; the urban areas support the opposition and the rural cast their votes for the ruling coalition. Without an insight into local politics, one could conclude that the Democratic Action Party (DAP), member of the anti-status-quo coalition, is an urban party, as all its elected representatives for the Sarawak state assembly are from urban constituencies.

A more careful look at the Sarawakian political scene would, however, reveal two important caveats: one is the differing levels of electoral competitiveness, with urban areas enjoying more political freedom, which makes urban areas more prone to vote for the opposition. DAP is the opposition party that fields candidates in urban seats in Sarawak. However, opposition parties are assigned seats according to the same ethnic pattern as it happens within the ruling coalition. Therefore, expanding beyond the party's traditional base is hampered by coalition partners and, for the most part, DAP – a traditionally Chinese party – is discouraged from mobilizing in the countryside, where there are few Chinese voters. There is, however, no indication of DAP trying to represent the urban residents per se. DAP is therefore a "Chinese" and "urban" party by the token of its elected representatives being Chinese and urban-based (particularly so in Sarawak); however, the party has so far contested only few and far apart rural and non-Chinese seats and its popularity in these areas cannot be ascertained, making it logically difficult to consider DAP an urban party.[20]

In West Kalimantan and Indonesia in general, the Democratic Party (PD) of former President Yudhoyono is a similar false positive as an urban party.[21] Its results in West Kalimantan demonstrate that it is much more popular in cities (Pontianak, Singkawang) than in rural areas. A mechanism similar to the one identified for DAP in Sarawak is in action here: the new Democratic Party competes chiefly against better established parties (Golkar and PDI-P) which enjoy entrenched support in rural and remote areas. As a new party, PD arrived with its message earlier in the cities, especially amongst the urban youth, and it will take time to establish electoral machinery strong enough to penetrate the remote rural areas. None of the parties however, PD or DAP, mobilizes along the "urban" affiliation and in fact both strive to make inroads in the interior of Sarawak and West Kalimantan, as this is their potential for growth. Therefore, neither of the two parties is ideologically an "urban" party.

The earlier caveat about the deeply penetrating patrimonialism suggests that class mobilization is unlikely in the cases analyzed here. The sets of "clients" and "patrons" may strike as corresponding to classes in Marx's sense, but, as Flynn (1974) showed, the behaviour of patrons and clients in politics against class mobilization is strikingly different:

(i)   The goals of the actors differ. In class or group politics, the parties' main aim is to translate their interests into general policy. Clientele politicians seek specific favours for their clients, in response to specific demands in a system where general ideological or programmatic formulations remain weak;

   (ii) The channels of communication are different. In group politics the politicians receive communications, both ideological and technological, from the class or interest group and, it is argued, are more open to general elite opinion;

   (iii) There is a difference in the locus of power. The clientelistic politician has more freedom, making only ad hoc arrangements with various groups, sometimes able to play off one against another and retain his autonomy.

(Flynn 1974, 138–139)

Scott is even more explicit in pointing out the irrelevance of class analysis in developing countries.

> [M]ost political groupings cut vertically across class lines and where even nominally class-based organizations like trade unions operate within parochial boundaries of ethnicity or religion or are simply personal vehicles. In a wider sense, too, the fact that class categories are not prominent in either oral or written political discourse in the Third World damages their a priori explanatory value.

(1972a, 91)

Therefore, the assumption is that ethnic mobilization coexists with intense patronage, nearly eliminating the possibility of class-based loyalties.

Another question that needs to be tackled is to what extent the national-level ideological rifts penetrate the regional-level politics. In both countries there is a clear political division between those who wish for a greater role of religion (Islam) in the state, and those who prefer a secular state (also identified by Lipset and Rokkan for Western European polities). If this issue is seen separately from the ethnic cleavage (Muslims are infinitely more likely to opt for Islamization than non-Muslims, although many Muslims support secularism), very few manifestations of this problem will be seen in regional politics. Political competition on the sub-national level revolves around questions of infrastructure, development and education, which often translate into very practical issues: allocations for schools in a particular region, a water treatment plant for a village, electricity connection for a constituency or tarring of a road from the interior to town. Political parties and their candidates, regardless of their party stands on broad ideological dilemmas, must engage in these mundane problems on the regional level. No meaningful ideological platform of parties or candidates was to be discerned from the political events I attended in both states: government and opposition rallies in Malaysia (2010) as well as public debates during the regent and gubernatorial election campaigns in West Kalimantan and PDI-P congress in Pontianak (2011).

In contemporary Malaysia, the anti-status-quo movement accounts for another electoral division. Opposition parties and civil society organizations call for ending the multi-decade-long rule of the National Front (Barisan Nasional, BN) coalition. The struggle against Barisan Nasional is waged under the banner of more transparency, freedom of speech and fully free political competition. This movement does not, however, seek to represent the interests of any particular group of voters. DAP, along with an Islamic party, PAS, and the main opposition party, People's Justice Party (Parti Keadilan Rakyat or PKR), currently act as a coalition challenging the BN in elections. All three are organizations attracting votes on the anti-status-quo platform, but each of them – if not bound by the coalition agreement – would likely opt for different solutions for the Malaysian state in the long term, whether in terms of state–religion relations, the economy or ethnic arrangements. Whether this movement can be recognized as a cleavage in the sense Lipset and Rokkan proposed is a matter of a more theoretically inclined work. Nevertheless, a study on electoral preferences must bear in mind this current strong political division in Malaysia. The opposition has so far been less successful in Sarawak than in West Malaysia.

To sum up, Indonesia and Malaysia are both countries in which patronage is rampant. Experience from other countries suggests that in patronage-driven contexts class mobilization is extremely difficult, while ethnic mobilization goes hand in hand with patronage. The assumption of ethnicity being a relevant cleavage is reinforced by the theory of visibility of ethnic identities versus relatively poorer visibility of other identities. Strictly ideological divisions in politics, I argue, are much less pronounced at the sub-national level, as regions are much more concerned about practical and material matters than ideological platforms. Consequently, we have shown that Sarawak and West Kalimantan are good laboratory settings to study ethnic identity change through the lens of political behaviour, as very few other political divisions are present in these cases.

1.5 Ethnic dimensions and categories in Malaysia and Indonesia

The main claim constructivists advocate about the nature of ethnic identity is that for each individual, there are multiple ethnic identities. Before I set off to study the shifts of ethnic identities, it is important to introduce the main dimensions of ethnic identities in the studied societies. The two provinces, Sarawak and West Kalimantan, share a very similar ethnic structure, even given the multiplicity of dimensions. Malaysia and Indonesia are both Muslim-majority states; however, the two regions discussed here are at best 50% Muslim, the remaining being on one hand, native tribes known by the umbrella exonym "Dayak" and on the other hand, "the Chinese", descendants of Chinese immigrants. Within Malaysia and Indonesia, this Malay-Dayak-Chinese threesome is particular to both Sarawak and West Kalimantan. However, these categories, albeit deeply rooted in the vernacular discourse, cannot be taken for granted. A short ethnographic overview of other ethnic categories in Bornean Sarawak and West Kalimantan is due.

"Malays", "Dayaks" and "Chinese" are categories most commonly referred to and discussed in vernacular terms, but empirically and historically informed catalogues are much more complex and involve not only many more categories, but also different dimensions of these categories. Without any pretence to create an

exhaustive listing, let us introduce the most obvious categories that have a potential to be activated, according to the dimensions along which they span:

- Indigeneity dimension: "Indonesian natives" and "Warga Negara Indonesia"
  ("Indonesian citizen"); "Bumiputera" ("sons of the soil") and "Chinese" and
  "Indian" in Malaysia.
- Religious dimension: Muslims, non-Muslims, Christians, Catholics, Protes-
  tants, animists, Buddhists, Confucianists.
- "Race" dimension: Malays, Chinese, Dayaks.
- "Tribal" dimension: Iban, Bidayuh, Orang Ulu, Kenayatn and hundreds more.
- Linguistic dimension: Malay, Melanau, Madurese, Javanese, Teochew,
  Hakka, Hokkien, Iban. There are dozens, if not hundreds, of other languages.
- Regional dimension: Sambas Malay, Pontianak Malay, Rajang Iban, Sari-
  bas Iban, Serian (Selako) Bidayuh, Bau-Lundu Bidayuh, West Kalimantan
  region, Kapuas Raya and many more.

These dimensions organize the two societies into any number of ethnic categories, between two (indigeneity dimension) and hundreds (linguistic dimension);
moreover, each individual belongs to one category on *each* dimension, and all
these categories together create her repertoire of ethnic identity categories. Therefore, she can be a Bumiputera Catholic Bidayuh speaker of the Bukar-Sadong
dialect from the Serian region in Sarawak. The empirically driven question of this
research reads therefore: which of these categories or combination of categories
are being activated in politics at any given time and why?

Political institutions and practice in both studied countries influence strongly
the data that is available in these states and information that can be drawn from
the data. In Malaysia, ethnic categories are sanctioned in the constitution and
in the party system and as such are explicitly and legally present in the political discourse; in Indonesia, not only ethnic parties are forbidden, but also ethnic
mobilization is officially frowned upon, therefore it becomes implicit. Explicit
versus implicit presence of ethnicity in public discourse has therefore influenced
the foci of analysis in both cases. This comparative study, while looking for ethnic
markers to watch for the change, will in both cases look in different places: in
the case of Malaysia the search will focus mostly on the explicit statements, only
in rare cases having to read between the lines. In Indonesia, mostly the indirect
statements will have a value for the research, while the explicit ones will be of
little use. Consequently, none of the data collected in the two countries is directly
comparable. There is neither a common measuring stick for the two cases, nor can
the particular types of data (press statements, interviews and quantitative data) be
compared across Malaysia and Indonesia. Only the final conclusions, drawn after
the data is analyzed, can be compared.

Chandra pointed out the main problem that emerges at the junction of the constructivist theory and the applicable methodology:

> [T]he implication for our data collection efforts is that they must make a dis-
> tinction between ethnic "structure" (the set of potential ethnic identities that

characterizes a population) and ethnic "practice" (the set of identities actually activated by that population), must accommodate the possibility of the multiplicity of identities in both structure and practice, and must be sensitive to context and time in collecting these data.

(2009, 2)

Simply put, most of the existing data sets include not only just one or two ethnic categories, but also categories included may not correspond to the ones that are actually politically activated, and hence, are of little interest to the researcher. Therefore, while working with the theory requires highly dynamic data sets, the primary quantitative data actually available for analysis is mostly of a static nature, i.e. censuses usually have predefined dimensions of ethnic identities, or even predefined categories within the dimensions. Therefore, an attempt to prove that different categories are activated during different elections/periods is often hampered by the lack of data, or the relevant information about potential activated categories has to be obtained from statistically less usable sources, e.g. estimates by parties and candidates, practical knowledge of local communities or historical information about migration dynamics. Moreover, the categories reflected in censuses are usually recognized for particular reasons or because of an agenda, and the census results have to be read with extensive knowledge of the local or national history and after taking into consideration the possible sensitivities and interests of the parties involved.

The underlying assumptions of this study indicate that knowing the exact ethnic/linguistic/religious distribution in a given constituency is crucial not only for researchers, but also for political entrepreneurs. As shown earlier, I assume that parties and political entrepreneurs may see fit to search for alternative ethnic categories in the population's repertoires that produce new minimum-winning coalitions. In order to do that, however, they need to possess knowledge of ethnic distribution. If parties and candidates truly did not know what the ethnic distribution pattern is, they would arguably give up the ethnic politics altogether, as strategic calculations of minimum-winning coalitions would be impossible. This is mere speculation: in reality, because of the visibility of ethnic markers, a rough estimation is always possible, even without an elaborate statistical approach. Moreover, commonsensical information is generally available: in Malaysia and Indonesia it does not take in-depth research to know that the Chinese are concentrated in urban areas, followers of Islam for historical reasons live along the coast, and Christian indigenous peoples are more likely to be found in the interior. Nevertheless, access to accurate, up-to-date information in the form of census tables is still in high demand among the politicians. Consequently, availability of information is of obvious political significance and this fact gives state agencies incentives to tamper with or obscure the data.

Malaysia has held censuses every 10 years since its independence. Census results are easily accessible, also on-line. The data include both ethnicity (categories: "Malay"; "Iban"; "Bidayuh"; "Melanau"; "Other Bumiputera"; "Chinese"; "Indian"; "Others") and religion (categories: Islam; Christianity; Confucianism/Taoism/tribal/folk/other traditional Chinese religions; other; no religion;

unknown). However, census data are aggregated by administrative district and offer little information about ethnic or religious distribution in electoral constituencies, as these are delineated independently from administrative boundaries. The Election Commission (EC) does collect its own ethnic data during voter registration, and EC's statistics are aggregated by constituency, but these statistics are not officially available. A few days before each election, the press usually obtains the data and publishes them. However, the EC does not entertain any requests by researchers to share the information.[22] Although the press reports only five categories (Malay/Melanau, Chinese, Iban, Bidayuh, Orang Ulu), other categories that are a majority or plurality in a given constituency are also reported on, and Chapter 3 will show several cases of categories like Kenayatn and Kayan, as well as Chinese and Bidayuh dialectical categories in this situation.

Indonesia has held censuses every 10 years since 1961, but until 2000 they did not include questions about ethnicity (Suryadinata, Arifin and Ananta 2003, xx). Data for this period is in the form of rough estimates based on the 1930 census. The newest sources of information about ethnic distribution in Indonesia are twofold: the censuses (2000 and 2010) (Badan Pusat Statistik 2000, 2010), which include the question of both ethnicity and religion, and the publications of the Ministry of Religious Affairs (Kementerian Agama 2010) which gathers its own data about distribution of religious followers in provinces, regencies and sub-districts (*kecamatan*). Both sources, the census and the ministerial data set, are suspected to be biased, as several of my interviewees indicated, and the data sets are somewhat inconsistent. I will, however, make use of both sources, as they complement each other.

The usefulness of the 2000 census (Badan Pusat Statistik 2000) data on ethnicity[23] in the case of West Kalimantan is highly questionable, regardless of the purpose of analysis. Some of the categories listed in the census are as good as irrelevant,[24] while the "Others" category accounts for the highest share of the population of the province. The absence of "Dayak" as a category is telling and suggests a major misrepresentation of the actual composition of the province. The 2010 census (Badan Pusat Statistik 2012) attempted to make good some mistakes from 2000, but was not able to avoid other mishaps. In both Indonesian censuses (2000 and 2010), respondents could choose from more than 100 categories of Dayaks alone; however, some categories were clustered into one (e.g. the Kenayatn category) and in the final tabulation came to significant numbers. Other categories were not clustered, albeit they are related to each other (e.g. Dayak Uud Danum and Dayak Uud Danum Cihie). Some categories were broken down into specific local distinctions, which made their numbers insignificant.

In practical terms, if one is trying to identify a political split within the Dayak category, the census-established categories are of little help. In one of the regencies (Landak), one category (Kenayatn) comprises about 60% of all Dayaks and 50% of the entire population of the *kabupaten* (regency). In another extreme case, *kabupaten* Sanggau, the biggest Dayak sub-category amounted to only 10% of all Dayaks and a mere 6% of the entire population of Sanggau. The misrepresentation may be also enhanced by the issue of multiple demonyms that many categories use. These account for many mistakes. The 2010 census found more than 60,000 "Dayak Pompang" in *kabupaten* Ketapang, although the Dayak Pompang are not mentioned in other sources, most notably in the most comprehensive book on Dayak sub-categories, locally compiled (Sujarni Alloy and Istiyani 2008), or the 2000 census. Therefore, the categorization of Dayak sub-categories and their numbers has to be treated with the utmost caution.

The 2010 census (Badan Pusat Statistik 2012) coded all Dayak sub-categories with the prefix "Dayak" to make sure that in the end all the non-Malay native peoples in Borneo can be put under one umbrella and the census-established number of Dayaks, regardless of sub-categories, reflects their actual strength in the province. This coding method was obviously designed to rectify the mistake from the 2000 census, where the "Dayak" category did not appear at all. However, in the 2010 census the categories of "Melayu Sambas" and "Melayu Pontianak" were coded with the prefix "Dayak". Because the two local categories of Malays amount to more than 600,000 people, the number of "Malays" (Melayu without prefix) is disproportionately low within the province (only 819,000), while Dayaks account for more than 2 million people because of the Sambas and Pontianak Malays who were classified as Dayaks. This suggests that the census may include more coding and/or tabulation mistakes.

This will be, however, of less concern to the current analysis than one might expect. According to an internal policy of the Statistics Office, the ethnicity-related census findings are not to be revealed for the level of district (*kecamatan*), i.e. are only available as aggregate data for *kabupaten* (regency) and province.[25] As it can be safely assumed, internal Dayak political divisions and cleavages between Muslim categories (Madurese, Javanese and Malays, due to their total numbers and distribution) can be only viable at the level of *kabupaten*, and in order to study them, one would need to analyze the ethnic composition at the level of *kecamatan*. If these are not available, such analysis based on the census becomes impossible. Similarly, the distinction between the Chinese of mixed parentage (*peranakan*) and those of Chinese-only background (also called *totok*) accounts for an ethnic division within the Chinese category; however, nothing is known about this division being activated in post–New Order politics,[26] and the census does not collect data allowing to test it in this book. Similarly, there is little information about Chinese language categories in West Kalimantan, and simply no trace of politically activated ones. The current project has no tools to test for the *totok-peranakan* or linguistic distinctions among the Chinese in electoral behaviour or in contemporary strategic mobilization in West Kalimantan.

Religious composition was the only data available that covers the entire province, all the *kabupaten* and *kecamatan*. The completeness of a religious data set obtained from the Ministry of Religion (Kementerian Agama 2010) gives this data set the obvious advantage over the census data: it allows a within-regency analysis. The Ministry of Religion compiles this data set annually, and results of the data are sometimes made available as part of popular Indonesian Statistics Office publications called ". . . dalam Angka" (or ". . . in Numbers"; e.g. "Kalimantan Barat dalam Angka"), issued for the country, all provinces and regencies.[27] However, it only captures the sole religious division, and the potential Malay-Javanese-Madurese cleavage cannot be appreciated with this data set, as these three categories share one religion. Similarly, divisions between Dayak sub-categories will not be visible through this data. Alternative information of ethnic composition of particular regions was available from secondary sources, and this will be used as much as possible to test if any divisions other than religion are activated in the society. A map of Dayak linguistic groups in West Kalimantan (Sujarni Alloy and Istiyani 2008) was useful to identify potential ethnic divisions within regencies, and in some cases these were confirmed through analysis of executive election results. Without this map no such analysis would be possible, just as tracing the cleavage between different categories among the Muslims could not be done. Simply, the distribution of the Javanese, Madurese, Bugis and Malays within regencies is unknown.

Figure 1.3 Dayak languages distribution in West Kalimantan

Source: *Peta Keberagaman Bahasa Dayak di Kalimantan Barat* (Sujarni Alloy and Istiyani 2008, CD).

To sum up, the analysis of the *kabupaten* (regency)-level elections, which are most likely to induce mobilization of alternative categories and therefore are potentially most interesting, is wanting with respect to certain ethnic categories. Data available are not detailed enough, campaigns are carried out locally and their coverage in the media is patchy. Information about alternative potential category dimensions (locality, language, religion) is unavailable to external observers, in respect both to candidates and to voters. Nevertheless, an attempt will be made to show how ethnic mobilization functions at this level.

The data availability difficulties in Indonesia mentioned earlier affect the analysis mostly because of the fact that candidates and parties are not willing to name the categories they wish to represent. This problem is absent in the analysis of ethnic mobilization in Sarawak – even in the absence of data, one can observe ethnic mobilization by following statements made by parties and candidates and political entrepreneurs who take advantage of permitted explicit mobilization. In Indonesia, limited data availability hampers the analysis as the researcher has to identify the potentially activated categories without them being named in the political discourse.

The next chapters are organized as follows: Chapter 2 presents the institutional developments in Sarawak and shows which categories were historically activated in the state, and through which institutions. Chapter 3 observes closely the past two cycles of electoral competition in Sarawak to test the extent to which the structure established in the first decades of Sarawak within Malaysia represents a fixed consociational pattern, and what elements in that pattern allow for more flexible mobilization of ethnic categories than what consociationalism envisions. Chapter 4 is again a historical overview, this time of West Kalimantan in Indonesia, with special attention being paid to the activation of ethnic categories in the province throughout the changing political environment. Chapter 5 looks at the modes of ethnic identity change in the new institutional setting (most importantly, the direct executive elections) and how it affects the historically relevant ethnic categories. Finally, Chapter 6 presents which institutions accounted for faster identity change in both countries, and which hamper the change, and based on that, some policy recommendations will be presented.

Notes

1 Under the primordialist umbrella, we can further distinguish between perennialists and essentialists, while among the early constructivists, situationalists and instrumentalists are to be differentiated, as shown in Wimmer (2013, chap. Introduction). Varshney (2007) offered a different categorization with interesting arguments against the view of instrumentalism being a subdivide of constructivism, and treating institutionalism as an approach on its own. Alternative to this interpretative divide, Brubaker (2009) presented a thorough categorization of approaches to ethnicity across fields and disciplines.

2 For a vernacular-terms rendition of the primordialism–constructivism dispute, compare Huntington's archetypically primordialist work *The Clash of Civilizations* (2003), and Amartya Sen's *Identity and Violence: The Illusion of Destiny* (2006), which argues the opposite in direct response to Huntington.

3 This person cannot change her identity to "Southerner", "Muslim" or "White" because her ancestry does not offer membership in these categories. Kanchan Chandra, who developed this pattern of ethnic change, called it "constrained change" (2008, 16–17) (more about this in Chapter 2).

4 Including religion under the ethnic umbrella is far from being uncontested, and throughout this book I recognize the difficulty of treating it as such, as religious adherence can be a choice and both cases studied here provide vivid examples of conversions, both mass and individual. Religious conversion is a starkly different form of "change" from the understanding of ethnic identity change as it is deployed here (see later in this chapter), and should not be confused. As this book shows, however, in many societies, including the two discussed here and, most notably, India, religion cannot be separated from ethnic identities, and "Malay" and "Hindu" are the best examples of it. Chandra (2004) and Varshney (2008) adopted a similar approach for their analyses of India.

5 Anderson (1983, chap. 10) recognized the importance of the census for identity building; for a discussion on census and identity, see Kertzer and Arel (2002). Hirschman (1987) analyzed ethnic categories of the census in Peninsular Malaysia, tracing them back to British rule.

6 For more sophisticated but less common electoral designs, like alternative vote (AV) and single transferable vote (STV), see Reilly (2002).

7 Consociationalism has been widely criticized, for several reasons: poor development of the theory (most notably Lustick 1997), the primordial understanding of ethnicity embedded in it, its undemocratic principle of an elite cartel that precludes actual choice in elections (Barry 1975), its applicability and suitability in particular cases where it was proposed (Horowitz 1991), and for the results it produced where implemented (Mehler 2009; Taylor 2006; Younis 2011).

8 Especially in the form of traditional laws that are recognized by the state and exist parallel to state laws. Also, Chinese and Tamils are allowed to run schools with Mandarin and Tamil as the medium of instruction. Note, however, that Malaysia's federalism cannot be seen as serving the principle of autonomy granted to ethnic categories, as not only the federal states were not designed as such, but also most states (Sarawak probably more than others) are too ethnically mixed to offer potential for substantial ethnic autonomy.

9 Donald Horowitz (1985) presented strong arguments against the claim that Malaysia is a consociational democracy, and he is correct that the Malaysian design falls short of fulfilling a couple critical requirements of consociationalism. However, we are concerned here with the question of whether and how ethnicity is embedded in institutions, and for the purpose of this question Malaysia can be considered "consociational", as it specifically names certain ethnic categories as entitled to share power or participate in politics. In other words, "consociationalism" is understood here as an epistemological tool to denote a political system that accommodates ethnic diversity by assigning some ethnic categories specific roles in the political system. I am indebted to an anonymous reviewer for raising this issue. For a summary of the debate on Malaysian consociationalism, see Davidson (2008b).

10 The other provinces being East Nusa Tenggara, Bali, Papua, West Papua, North Sulawesi and Maluku.

11 82% of the province's population uses some local dialect or language (*bahasa daerah*) for their daily communication (Na'im and Syaputra 2012, 48).

12 Compare Peters (2004, chap. 4) for discussion of the important problem of comparability of concepts and measurements.

13 Davidson (2008a) and Tanasaldy (2012) (the latter implicitly) applied historical institutionalism in their studies of West Kalimantan; Davidson to show the perpetuity of ethnically framed riots, Tanasaldy to show the differing fate of an ethnic category in the changing political environment.

14 Compare Cox (1997) and Shepsle, Rhodes and Godin (2006) for discussion on definitions of institutions with a focus on their endo- and exogeneity.

15 For a detailed study of incremental institutional changes during the Indonesian New Order, see Slater (2010), and for shifts of power also outside of institutions in that time in Indonesia, see Ufen (2002).

16 It has to be kept in mind that Indonesia is an atypical case of ethnic party bans, as it allows religious parties, but bans regional parties (except for Aceh province), eliminating local ethnic parties. Parties appealing to ethnic categories can fulfil statutory requirements to operate if the category is dispersed across all provinces in Indonesia (take the Chinese and, albeit ephemeral, Chinese parties), but parties mobilizing Hindu followers or Dayaks (both categories are geographically concentrated in only one or two provinces) are not permitted.

17 Although rational choice gave an intellectual impulse for this research, there is no ambition here to work with formal modelling. The background to the rational choice line of thought was acquired from Von Neumann and Morgenstern (1944), an all-time must-read in the field of game theory; McCarty and Meirowitz (2007) served to assess the potential of formal model-based study for the questions posed here.

18 For wider discussion of this issue see Means (1991), Case (1996), Crouch (1996) and Ufen (2008b).

19 Discussing oligarchization of elections in Indonesia, Mietzner comes to similar conclusions (2013, chap. 5).

20 The absence of rural-urban political division in Sarawak was first pointed out by Leigh (1979).

21 Although PD is a non-ideological party, it was observed that it primarily targets the urban electorate; compare Ananta, Arifin and Suryadinata (2005). The Prosperous Justice Party (Parti Keadilan Sejahtera or PKS) is not only an Islamic party, but one that also targets urban electorates. Across Indonesia its popularity was highest in cities, in particular in Jakarta, and the same trend was visible in West Kalimantan. In the city of Pontianak PKS obtained 10% of votes, against its 4% of support in the province on average.

22 I received no reply from EC to my written request.

23 For a discussion of the construction of the 2000 census and the questionnaire, as well as methodological mishaps, see Suryadinata, Arifin and Ananta (2003).

24 Compare Hidayah (1996), who listed ethnic categories in West Kalimantan different from those accounted for in the census.

25 Email communication with the Statistics Office (Badan Pusat Statistik) from 17 October 2012. Upon my request for the raw, untabulated data for the relevant variables ("ethnic group", "language spoken at home", "religion"), I was informed that the raw data would not include coding allowing me to identify the district.

26 For the post-independence situation of the Chinese community and the then-prominent *totok-peranakan* division, see Suryadinata (1972), Mackie (1976), Siddique and Suryadinata (1981), Suryadinata (1997), Hui (2011). Dawis (2009) explores the dynamic cultural context of the Chinese in Indonesia and the transition from the Chinese language ban under Suharto to the relative embracement of multiculturalism during Reformasi. Dawis argues that the *totok-peranakan* division may have long lost its relevance (2009, 80–81).

27 Indeed, the publication uses the Ministry of Religion data set for the district levels, although the publisher, the Statistics Office, gathers the same data in the census.

References

Ananta, Aris, Evi Nurvidya Arifin and Leo Suryadinata. 2005. *Emerging Democracy in Indonesia*. Singapore: ISEAS.

Anderson, Benedict. 1983. *Imagined Communities: Reflections on the Origin and Spread of Nationalism*. London: Verso.

Badan Pusat Statistik. 2000. *Kalimantan Barat Dalam Angka 2000*. Pontianak: BPS.

———. 2010. *Penduduk Menurut Wilayah dan Agama yang Dianut. Provinsi Kalimantan Barat*. Badan Pusat Statistik Online. www.sp2010.bps.go.id.

———. 2012. *Suku Bangsa. Sensus Penduduk 2010*. N.p. Unpublished.

Barry, Brian. 1975. "The Consociational Model and Its Dangers". *European Journal of Political Research* 3 (4): 393–412.

Barth, Fredrik, ed. 1970. *Ethnic Groups and Boundaries*. Reprint. Bergen [u.a.]: Univ.-Forl. [u.a.].

Bates, Robert H. 1974. "Ethnic Competition and Modernization in Contemporary Africa". *Comparative Political Studies* 6 (4): 457–484.

Bogaards, Matthijs. 2003. "Electoral Choices for Divided Societies: Multi-Ethnic Parties and Constituency Pooling in Africa". *Commonwealth & Comparative Politics* 41 (November): 59–80.

Brubaker, Rogers. 1996. *Nationalism Reframed*. Cambridge: Cambridge University Press.

———. 2002. "Ethnicity without Groups". *European Journal of Sociology / Archives Européennes de Sociologie* 43 (02): 163–189.

———. 2004. *Ethnicity without Groups*. Cambridge, MA: Harvard University Press.

———. 2009. "Ethnicity, Race, and Nationalism". *Annual Review of Sociology* 35 (1): 21–42.

Brubaker, Rogers, and David Laitin. 1998. "Ethnic and Nationalist Violence". *Annual Review of Sociology* 24: 423–452.

Case, William. 1996. *Elites and Regimes in Malaysia*. Monash Papers on Southeast Asia. Clayton: Monash Asia Institute.

———. 2011. *Executive Accountability in Southeast Asia: The Role of Legislatures in New Democracies and under Electoral Authoritarianism*. Policy Studies 57. Honolulu: East-West Center.

Chandra, Kanchan. 2004. *Why Ethnic Parties Succeed: Patronage and Ethnic Head Counts in India*. Cambridge Studies in Comparative Politics. Cambridge, UK ; New York: Cambridge University Press.

———. 2008. "Ethnic Invention: A New Principle for Institutional Design in Ethnically Divided Democracies". In *Designing Democratic Government: Making Institutions Work*, edited by Margaret Levi, James Johnson, Jack Knight and Susan Stokes, 89–115. New York: Russell Sage Foundation.

———. 2009. "A Constructivist Dataset on Ethnicity and Institutions (CDEI)". In *Identity as a Variable: A Guide to Conceptualization and Measurement of Identity*, edited by Rawi Abdelal, Yoshiko M. Herrera, Alastair Iain Johnston and Rose McDermott. Cambridge: Cambridge University Press.

———. 2011. "What Is an Ethnic Party?" *Party Politics* 17 (2): 151–169.

———. 2012a. "Attributes and Categories: A New Conceptual Vocabulary for Thinking about Ethnic Identity". In *Constructivist Theories of Ethnic Politics*, 97–131. Oxford: Oxford University Press.

———., ed. 2012b. *Constructivist Theories of Ethnic Politics*. Oxford: Oxford University Press.

———. 2012c. "How Ethnic Identities Change". In *Constructivist Theories of Ethnic Politics*, 132–178. Oxford: Oxford University Press.

———. 2012d. "What Is Ethnic Identity? A Minimalist Definition". In *Constructivist Theories of Ethnic Politics*, 51–96. Oxford: Oxford University Press.

Chin, James. 1996. *Chinese Politics in Sarawak: A Study of the Sarawak United People's Party*. South-East Asian Social Science Monographs. Kuala Lumpur, Malaysia and New York: Oxford University Press.

Collier, David, Henry E. Brady and Jason Seawright. 2004. "Sources of Leverage in Causal Inference: Toward an Alternative View of Methodology". In *Rethinking Social Inquiry: Diverse Tools, Shared Standards*, edited by Henry E. Brady and David Collier, 161–200. Lanham, MD: Rowman & Littlefield.

Cox, Gary W. 1997. *Making Votes Count*. Cambridge: Cambridge University Press.

———. 1999. "Electoral Rules and Electoral Coordination". *Annual Review of Political Science* 2 (1): 145–161.

Crouch, Harold. 1996. *Government and Society in Malaysia*. Ithaca, NY and London: Cornell University Press.

Davidson, Jamie S. 2008a. *From Rebellion to Riots: Collective Violence on Indonesian Borneo*. New Perspectives in Southeast Asian Studies. Madison: University of Wisconsin Press.

———. 2008b. "The Study of Political Ethnicity in Southeast Asia". In *Southeast Asia in Political Science: Theory, Region and Qualitative Analysis*, edited by Erik Martinez Kuhonta, Dan Slater and Tuong Vu, 199–226. Stanford, CA: Stanford University Press.

Dawis, Aimee. 2009. *The Chinese of Indonesia and Their Search for Identity: The Relationship between Collective Memory and the Media*. Amherst, NY: Cambria Press.

Fearon, James. 1999. "Why Ethnic Politics and 'Pork' Tend to Go Together." Presented at an SSRC-MacArthur sponsored conference on "Ethnic Politics and Democratic Stability," University of Chicago, May 21–23, 1999. www.stanford.edu/group/fearon-research/cgi-bin/wordpress/wp-content/uploads/2013/10/Pork.pdf

Flynn, Peter. 1974. "Class, Clientelism, and Coercion: Some Mechanisms of Internal Dependency and Control". *Commonwealth & Comparative Politics* 12 (2): 133–156.

Hadiz, Vedi R. 2010. *Localising Power in Post-Authoritarian Indonesia: A Southeast Asia Perspective*. Stanford, CA: Stanford University Press.

Hazis, Faisal S. 2012. *Domination and Contestation: Muslim Bumiputera Politics in Sarawak*. Singapore: ISEAS.

Hidayah, Zulyani. 1996. *Ensiklopedi Suku Bangsa Di Indonesia*. Jakarta: LP3ES.

Hirschman, Charles. 1987. "The Meaning and Measurement of Ethnicity in Malaysia: An Analysis of Census Classifications". *Journal of Asian Studies* 46 (3): 555–582.

Horowitz, Donald. 1985. *Ethnic Groups in Conflict*. Berkeley: University of California Press.

———. 1991. *A Democratic South Africa? Constitutional Engineering in a Divided Society*. Perspectives on Southern Africa. Berkeley [u.a.]: University of California Press.

———. 2013. *Constitutional Change and Democracy in Indonesia*. Cambridge: Cambridge University Press.

Hui, Yew-Foong. 2011. *Strangers at Home: History and Subjectivity among the Chinese Communities of West Kalimantan, Indonesia*. Leiden and Boston, MA: BRILL.

Huntington, Samuel P. 2003. *The Clash of Civilizations and the Remaking of World Order*. New York: Simon & Schuster.

Indrayana, Denny. 2008. *Indonesian Constitutional Reform, 1999–2002: An Evaluation of Constitution-Making in Transition*. Jakarta: Penerbit Kompas Buku.

Jayum, A. Jawan. 1991. *The Ethnic Factor in Modern Politics: The Case of Sarawak, East Malaysia*. Occasional Paper, no. 20. Hull: Centre for South-East Asian Studies.

Katz, Richard S., and Peter Mair. 2009. "The Cartel Party Thesis: A Restatement". *Perspectives on Politics* 7 (04): 753–766.

Kementerian, Agama. 2010. *Jumlah Penduduk Menurut Agama per Kecamatan Propinsi Kalimantan Barat Tahun 2010*. N.p. Unpublished.

Kertzer, David I., and Dominique Arel. 2002. *Census and Identity: The Politics of Race, Ethnicity, and Language in National Census*. Cambridge, UK; New York: Cambridge University Press.

Leigh, Michael B. 1974. *The Rising Moon*. Sydney: Sydney University Press.

———. 1979. "Is There Development in Sarawak? Political Goals and Practice". In *Issues in Malaysian Development*, edited by James C. Jackson and Martin Rudner, 339–374. Singapore: Heinemann Educational Books.

Lijphart, Arend. 1969. "Consociational Democracy". *World Politics* 21 (2): 207–225.

———. 2004. "Constitutional Design for Divided Societies". *Journal of Democracy* 15 (2): 96–109.

Lipset, Martin Seymour, and Stein Rokkan. 1967. *Party Systems and Voter Alignments: Cross-National Perspectives*. New York: Free Press.

Lustick, Ian S. 1997. "Lijphart, Lakatos, and Consociationalism". *World Politics* 50 (01): 88–117.

Mackie, J. 1976. *The Chinese in Indonesia : Five Essays*. Honolulu: University Press of Hawaii in association with Australian Institute of International Affairs.

Mahoney, James. 2000. "Rational Choice Theory and the Comparative Method: An Emerging Synthesis?" *Studies in Comparative International Development* Summer 2000: 111–141.

McCarty, Nolan, and Adam Meirowitz. 2007. *Political Game Theory: An Introduction*. Cambridge: Cambridge University Press.

Means, Gordon P. 1991. *Malaysian Politics: The Second Generation*. Singapore: Oxford University Press.

Mehler, Andreas. 2009. "Peace and Power Sharing in Africa: A Not So Obvious Relationship". *African Affairs* 108 (432): 453–473.

Mietzner, Marcus. 2013. *Money, Power, and Ideology: Political Parties in Post-Authoritarian Indonesia*. Honolulu: University of Hawaii Press.

Mozaffar, Shaheen, James R. Scarritt and Glen Galaich. 2003. "Electoral Institutions, Ethnopolitical Cleavages, and Party Systems in Africa's Emerging Democracies". *American Political Science Review* 97 (03): 379–390.

Na'im, Akhsan, and Hendry Syaputra. 2012. *Kewarganegaran, Suku Bangsa, Agama Dan Bahasa: Hasil Sensus Penduduk 2010*. Jakarta: Badan Pusat Statistik.

Peters, Brainard Guy. 2004. *Comparative Political Analysis*. Hagen: Fernuniversität in Hagen.

Pierson, Paul, and Theda Skocpol. 2002. "Historical Institutionalism in Contemporary Political Science". In *Political Science: The State of the Discipline*, edited by Ira Katznelson and Helen V. Milner. New York: W. W. Norton & Co.

Posner, Daniel. 2005. *Institutions and Ethnic Politics in Africa*. Cambridge: Cambridge University Press.

Przeworski, Adam. 2007. "Capitalism, Democracy and Science". In *Passion, Craft, and Method in Comparative Politics*, edited by Gerardo L. Munck and Richard Snyder, 456–503. Baltimore, MD: Johns Hopkins University Press.

Przeworski, Adam, and Henry Teune. 1970. *The Logic of Comparative Social Inquiry*. New York: Wiley-Interscience.

Rabushka, Alvin, and Kenneth A. Shepsle. 2009. *Politics in Plural Societies*. New York: Pearson/Longman.

Reilly, Benjamin. 2002. "Electoral Systems for Divided Societies". *Journal of Democracy* 13 (2): 156–170.

———. 2006. "Political Engineering and Party Politics in Conflict-Prone Societies". *Democratization* 13 (5): 811–827.

———. 2011. "Centripetalism". In *Conflict Resolution: Theory and Practice*. London: Routledge. http://ethnopolitics.org/isa/Reilly.pdf.

Riker, William H. 1962. *The Theory of Political Coalitions*. New Haven, CT: Yale University Press.

———. 1986. *The Art of Political Manipulation*. New Haven, CT: Yale University Press.

Scott, James C. 1972a. "Patron-Client Politics and Political Change in Southeast Asia". *The American Political Science Review* 66 (1): 91–113.

———. 1972b. "The Erosion of Patron-Client Bonds and Social Change in Rural Southeast Asia". *The Journal of Asian Studies* 32 (1): 5–37.

Searle, Peter. 1983. *Politics in Sarawak, 1970–1976: The Iban Perspective*. Singapore ; New York: Oxford University Press.

Sen, Amartya. 2006. *Identity and Violence: The Illusion of Destiny*. New York: W. W. Norton & Co.

Shepsle, Kenneth A. 2006. "Rational Choice Institutionalism." In *Oxford Handbook of Political Institutions*, edited by R. A. W. Rhodes, Sarah A. Binder and Bert A. Rockman, 23–38. Oxford; New York: Oxford University Press.

Siddique, Sharon, and Leo Suryadinata. 1981. "Bumiputra and Pribumi: Economic Nationalism (Indiginism) in Malaysia and Indonesia". *Pacific Affairs* 54 (4): 662–687.

Slater, Dan. 2004. "Indonesia's Accountability Trap: Party Cartels and Presidential Power after Democratic Transition". *Indonesia* 78 (October): 61–92.

———. 2010. "Altering Authoritarianism: Institutional Complexity and Autocratic Agency in Indonesia". In *Explaining Institutional Change: Ambiguity, Agency and Power*, edited by James Mahoney and Kathleen Thelen 132–167. Cambridge: Cambridge University Press.

Subianto, Benny. 2009. "Ethnic Politics and the Rise of the Dayak Bureaucrats in Local Elections: Pilkada in Six Kabupaten in West Kalimantan". In *Deepening Democracy in Indonesia?: Direct Elections for Local Leaders (Pilkada)*, edited by Maribeth Erb and Priyambudi Sulistiyanto, 327–351. Singapore: ISEAS.

Sujarni Alloy, Albertus, and Chatarina Pancer Istiyani. 2008. *Keberagaman Subsuku Dan Bahasa Dayak Di Kalimantan Barat*. Edited by John Bamba. Pontianak: Institut Dayakologi.

Suryadinata, Leo. 1972. "Indonesian Chinese Education: Past and Present". *Indonesia*, no. 14: 49–71.

———. 1997. *Political Thinking of the Indonesian Chinese, 1900–1995: A Sourcebook*. Singapore: Singapore University Press, National University of Singapore.

Suryadinata, Leo, Evi Nurvidya Arifin and Aris Ananta. 2003. *Indonesia's Population: Ethnicity and Religion in a Changing Political Landscape*. Singapore: ISEAS.

Tanasaldy, Taufiq. 2012. *Regime Change and Ethnic Politics in Indonesia: Dayak Politics in West Kalimantan*. Leiden: KITLV.

Taylor, Rupert. 2006. "The Belfast Agreement and the Politics of Consociationalism: A Critique". *The Political Quarterly* 77 (2): 217–226.

Tilly, Charles. 2008. *Democracy*. Cambridge: Cambridge University Press.

Ufen, Andreas. 2002. *Herrschaftsfiguration Und Demokratisierung in Indonesien (1965–2000)*. Mitteilungen Des Instituts Für Asienkunde, Hamburg. Hamburg: IFA.

————. 2008a. "From Aliran to Dealignment: Political Parties in Post-Suharto Indonesia". *South East Asia Research* 16 (1): 5–41.

————. 2008b. "The 2008 Elections in Malaysia: Uncertainties of Electoral Authoritarianism". *Taiwan Journal of Democracy* 4 (1): 155–169.

Varshney, Ashutosh. 2007. "Ethnicity and Ethnic Conflict". In *The Oxford Handbook of Comparative Politics*, 274–94. Oxford: Oxford University Press.

————. 2008. *Ethnic Conflict and Civic Life: Hindus and Muslims in India*. New Haven, CT: Yale University Press.

Von Neumann, John, and Oskar Morgenstern. 1944. *Theory of Games and Economic Behavior*. Princeton, NJ: Princeton University Press.

Wimmer, Andreas. 2013. *Ethnic Boundary Making: Institutions, Power, Networks*. 1 edition. New York: Oxford University Press.

Younis, Nussaibah. 2011. "Set up to Fail: Consociational Political Structures in Post-War Iraq, 2003–2010". *Contemporary Arab Affairs* 4 (1): 1–18.

2 Sarawak
Institutional and historical overview

2.1 Consociation and ethnic categories

As was indicated in Chapter 1, the Malaysian and Sarawakian political institutions and political practice revolve around the consociational paradigm, i.e. based
on an elite bargain, particular ethnic categories are legitimately invoked, repeat-
edly mobilized and participate in government according to a more or less explicit
power-sharing agreement. By principles of consociationalism, this type of polity
cannot live without fixed and steady, census-measurable categories. Therefore,
the fixedness and steadiness of ethnic practice is a theory-drawn hypothesis for
consociational societies, including Sarawak.

There are, however, important caveats to the consociational assertion; foremost,
if "ethnic groups" are to share power, as we came to believe is the case in Malaysia, how is it decided which "groups" are entitled to share some piece of the power
cake? In many societies the cleavages may be obvious enough, as seems to have
been the case in Malaya, where the "Malays", "Chinese" and "Indians" are commonsensically distinguishable and visible with the naked eye.[1] But in Sarawak
there is nothing obvious about the ethnic cleavages, or at least there was not at
the time when the power-sharing scheme was imposed on the state. Therefore, we
will be asking: how were the currently activated categories developed to become
the *titular categories*, or the categories entitled to share the power? Is there indeed
a fixed set of titular categories? Are they represented by *titular parties*? If power
is shared between these titular categories, do people retain any other activated
categories in their repertoires? How is it achieved? Resolving these issues is a
necessary step to arrive at the final problem of this thesis: does consociationalism
invariably arrest ethnic identities?

To answer the compelling question of the establishment of titular categories,
i.e. those of the categories that share the power, we will invoke historical institutionalism and retrace the steps of political parties, the central government in
Kuala Lumpur and prominent Sarawakian leaders to see how the categories were
invented and re-invented to serve the purpose of power sharing. Critically, initial
political mobilization in Sarawak happened before the state joined the Federation
and before the incentives of the consociational institutions were known to the
players. The first election and the first coalition negotiations took place within an

institutional setting that was very different from the Malaya-style politics introduced later. The subsequent power-sharing bargain was established within institutions transplanted along with the constitution and coalition governments (under
the banner of Alliance or National Front, or Barisan Nasional) from Malaya. As
it turned out, the power was to be shared not only in the executive and legislative
realms, but also on the national and state levels and among parties and elites. The
constitutionally designed institutions formally regulate the political process; however, due attention should be paid to practices that go beyond the legal regulations.

Therefore, armoured with the political coalitions theory (Riker 1962), we will
attempt to define the criteria established in Sarawak for winning coalitions (except
for the obvious majority in the legislature); we will look out for sequences of joining the coalition to establish the practical strength of each component and possible
privileges that certain components may have appreciated over time and, alternatively, which components were smaller in size but pivotal in decision making, and
which were circumvented despite their numerical strength. Consociationalism
assumes representation of all, i.e. each and every citizen must be a member in at
least one of the categories entitled to share the power cake. In Sarawak this was
not necessarily always how leaders or parties imagined it. The very first coalitions
created in Sarawak in the 1960s, before the Malaya-style politics took over, were
coalitions of exclusion. Most parties had a clear preference as to the quarters of
society with which they wished to cooperate and with which they did not. The
West Malaysian model, however, required a transition to a multiethnic coalition
that involved representation of all (at the time relevant) ethnic categories. Ideally,
each titular category should be represented by one party in the coalition – that was
at least the model the West Malaysian powers-that-be strove for.

The process of co-opting parties without which the Sarawakian ruling coalition
would not quite have been an all-inclusive one is an important part of this chapter.
Equally important, in the course of events, parties were eliminated and replaced
within the coalition. However, highest attention was paid to preventing any distor-
tion of the ethnic balance in the government or weakening of its legitimacy, which
has always been pegged to the idea of ethnic compromise and cooperation. All
this is to say, the historical trajectory of Sarawakian party politics and leadership
may indicate crucial characteristics of methods of manipulating consociational
politics in the particular institutional arena.

After following the historic trajectories of manoeuvring of ethnic categories,
first by parties, subsequently by the ruling coalition, we will arrive at the current
political happenings. The main question the latter sections are to answer is: how
is it assured that each titular ethnic category is given exactly the proportion of
power that keeps it satisfied, and the others remain content? The question is not
necessarily "how much power", but rather "in what form" is the power shared to
give the proper image of proportionality. Is it the number of ministerial portfolios
in federal and state cabinets? Is it the number of representatives each category
elects, assuming the categories elect their co-ethnics? Or is it the number of constituencies in which the category constitutes a majority? Finally, who is the agent
who holds the power in the name of a given category? Is it the ethnic parties that

(claim to) represent particular ethnic categories and participate in government in their name?

Analysis of this consociational polity indicates a much more dynamic pattern of ethnic identity change than the theory of consociationalism would indicate. This dynamism will be attributed partly to chronological changes of institutions and incentives, and partly to the fact that elite bargaining in Sarawak happens through at least three channels: political parties, legislative seats and executive (non-elected) positions. Multiple channels of power sharing allow for the activation of multiple sets of categories in the society, as will be observed in Sarawak. Therefore, this research finds not only significant temporal changes in ethnic identities, but also frequent shifts between categories in an individual's repertoire within a short period, depending on political context.

The analysis of the sequence of ethnic category activation in Sarawak starts with the beginning of electoral politics in the state, which chronologically coincides with the British decision to withdraw from the region. It is of advantage to frame the analysis on a chronological axis and pay attention to the early years of Sarawakian politics, which, given strong links of continuity in state political life, will prove the origin and/or cause of many later developments. No new research is provided for the period 1960–2000. Secondary sources, supported by election results from the Electoral Commission, will be utilized here to discern the changing ethnic identities and their relevance. Therefore, we will find the origins of most of the currently relevant categories in the 1960s and 1970s parties' establishment and evolution. It is therefore important to look at these early cleavages and their cross-cuts to see how they evolved and achieved the current form. Sarawak's ethnic composition was "almost terrifying in its complexity" (Milne 1967, 51) at the beginning of the period studied here, and this statement holds true until the current day.

The currently most common tripartite division of Muslim indigenous-non-Muslim indigenous-Chinese was not always taken for granted. As Roff observed, "the political leaders of East Malaysia have done much by way of defining (and redefining) the groups they purported to represent, and in the process raised their consciousness and pride in cultural distinctiveness" (1974, 9), unequivocally suggesting the fluidity of ethnic loyalties. Leigh called the simplistic but widespread tripartite image of the Sarawakian political scene a "communal caricature of political recruitment [that] is very far from watertight and is quite inadequate as an explanation of political behavior" (1974, 203). To support his point, Leigh shows further evidence of existing cross-cleavages (education, class, profession etc.) among the three categories, thus proving that it is a gross oversimplification to see the politics as tripartite. Here, without resorting to any other cleavages, which in this work are assumed to be constant and of weak explanatory value, I will look for evidence of the activation of other and multiple *ethnic* categories and a more dynamic picture of the ethnic politics than the one many scholars envisioned.

In this work, I will continue to speak of "non-Muslim indigenous" or "non-Muslim natives" when referring to the vernacular meaning of "Dayak", while "Dayak" in this text is reserved for the politically activated category.

While in the 1960s and 1970s "non-Muslim native politics" was indeed equivalent to "Iban politics" ("Iban" being only one among several other categories included within "Dayak"), because of the lack of political organization among other non-Muslim indigenous, later it was much less so. By the 1980s, non-Iban categories among the natives caught up politically. Constituencies in which the Bidayuh and other indigenous were the majority became important battlefields in elections and the Bidayuh and Orang Ulu became categories of politics. However, by then the term "Dayak" was also activated in an attempt to unite all non-Muslim indigenous under its umbrella.

In some texts "non-Muslim Bumiputera"[2] is used to avoid the misleading and multi-meaning "Dayak". Here this term will be avoided, as "Bumiputera" is an ethnic category that, although so far never successfully activated, was introduced to Sarawak along with the West Malaysian political division and has a political value. The term "Bumiputera" is not used in the constitution itself but is in common use not only in informal language, but also in documents issued by state institutions. Its relevance is enormous because of affirmative action policies that have been in operation in Malaysia since independence. The constitution (art. 153) guarantees that "the special position of Malays and natives" is "safeguarded". Further protective measures were introduced in 1971 with the New Economic Policy (NEP). NEP and its later forms[3] are Malaysian affirmative action policies oriented at the alleviation of economic disparities between wealthy Chinese and Indian communities and the rest of the society. NEP sanctioned a wide range of privileges to peoples native to Malaysia, including a quota in university admission and civil service, granting government contracts to companies owned by Bumiputeras, better access to loans etc. To qualify for membership in the Bumiputera category, one has to have at least one parent who belongs to any of the constitutionally defined categories.[4] In practical terms, in Sarawak it is the Chinese who are excluded from the Bumiputera category.

Siddique and Suryadinata observe that before 1981,

> "Bumiputera" was not used in official government documents as a collective term to refer to both Malays and indigenous people of Sabah and Sarawak. In the Second and Third Malaysia Plans, for example, the phrase "Malays and other indigenous people" is used. It is of some significance, therefore, that in the newly issued Fourth Malaysia Plan [1981–1985], the term "Bumiputera" peppers the text and, although no official definition is provided, it is used in contexts where "Malays and other indigenous people" would previously have appeared.
>
> (1981, 674)

Note, however, that Bumiputera is hardly a vernacular category; on the contrary, it was specifically invented as a vessel to carry West Malaysian Malays and both East Malaysian Malays and other indigenous groups. An epistemological trouble at any rate, ontologically "Bumiputera" is quite clear: all non-Chinese in Sarawak are Bumiputera (Sarawak has negligible numbers of Indians).

A greater terminological difficulty is attached to the terms "Bidayuh" and "Orang Ulu". Both of these refer to everyday use as well as political categories and there are no apolitical equivalents to them. The people now known as Orang Ulu, prior to their activation in politics, used to be referred to as "other indigenous", which is as useless a term to a researcher as it is to a politician. The membership rule for "Orang Ulu" is chiefly the region of origin – upstream, far interior areas of Sarawak, mostly along the Indonesian border. The category comprises many linguistic categories; listing all possible tribal categories that comprise the Orang Ulu is simply impractical and can hardly do justice to all. The Bidayuh could be referred to by the constitutional term "Land Dayaks", but the term being rejected in the vernacular, praxis and politics for decades would now appear obsolete and out of touch with reality. In sum, as a general rule, these ethnonyms in the text will refer to activated ethnic categories.

"Muslim" is not an activated category in Sarawak. Being a Muslim, however, is a membership rule for the Malay, Melanau and Kedayan categories. Especially the latter two would otherwise be also eligible for membership in the Dayak category, as they are indigenous converts to Islam, but Sarawak's ethnic practice excludes Muslims from the Dayak category. Kedayan is too small in numbers to be an object of day-to-day political dealings, but the other two categories came to be known as "Malay/Melanau", or sometimes Muslim Bumiputera. Malays and Melanaus, however, are geographically separated and one's place of origin and language are clear distinctions allowing for Malay, Melanau and Kedayan to retain each of these categories in one's repertoire, next to Muslim Bumiputera.

"Chinese" is a more straightforward category: the basic membership rule is "descendants of immigrants from China". Religion does not play a role in this category, although Chinese converts to Islam are in a sort of gray zone; as speakers of the Malay language and Muslims they can claim membership in the Malay category. These cases are, however, very rare. While the other religious divisions are not activated among the Chinese (i.e. politically it is not divisive whether they are Catholics, Protestants, Buddhists or Confucianists), the linguistic or dialectical divisions do matter. Many Chinese may not speak their ancestors' language, and may not be educated in Mandarin, but their belonging to one of the "clans" or dialectical categories (Hakka, Hokkien, Teochew etc.) may still be of political relevance.

2.2 Sarawak within the Federation of Malaysia

Malaya's constitutional framework of citizenship and the legal position of ethnic groups was decided before the state formally came into existence in 1957. Before the British ceded power and granted independence to Malaya, a commission (called the Reid Commission after the name of its chair) was established in 1956 to investigate preferences of the different "races" (as Malaysians commonly refer to their different ethnic categories), the Malays, Chinese and Indians, towards the institutional shape of the future independent state in respect to rights and privileges of the different ethnic components. The commission was also to advise on a constitutional solution for the Federation. Based on the findings of the commission a compromise was struck between the economically underprivileged Malays and the better-off Chinese and Indians.

Although the quid pro quo is not specifically mentioned in the constitution, its terms are reflected in articles 14 to 18 pertaining to citizenship rights, which were granted irrespective of race to all residents of Malaya, and article 153, which guarantees privileges for Malays. According to the constitution (art. 153), Yang di-Pertuan Agong (i.e. the monarch, elected from among the Malay rulers) "is also obliged to safeguard the special position of the Malays". This "special position" is further explained in the article as the reservation of positions in public service, scholarships, permits or licenses for the operation of any trade or business that requires a permit or license. Moreover, the symbols of the Federation of Malaysia are designed according to Malay traditional symbols; also the national language (Malay), religion (Islam)[5] and the position of the sultans and the Yang di-Pertuan Agong all reflect the "Malayness" of the state.

The creation of the Federation of Malaysia blurred this clear-cut vision of the Malay state. As the British decided to withdraw from Singapore, Sarawak and North Borneo (now called Sabah) in the early 1960s, they proposed a merger of the three entities within the Federation of Malaysia.[6] A commission was established to look into the peoples' preferences in the two Bornean states (Sarawak and Sabah) relating to their political future. The commission's findings were inconclusive, as it found that the Malays supported the merger and the Chinese were against it, while the non-Muslim natives were not sufficiently consulted to make an informed decision.[7] An Iban journalist put it this way: "We were forced to join Malaysia, although we knew it was no good for us"[8] Nevertheless, the merger was finalized in 1963.

The constitution was now amended to accommodate the new members of the Federation; the state was to be called the Federation of Malaysia, or Malaysia; Malay privileges were to be extended to all indigenous peoples in Sarawak and Sabah (most of which in the two states were non-Muslims). A period of 10 years was guaranteed until full transition from the English language to Malay in the new territories; religious provisions remained unchanged in reference to the Federation, but Islam was not to be the religion of the new states because of Muslims being a minority in these states (Milne 1967, 69). Notably, without Sarawak and Sabah, the Singaporean Chinese would have shifted the Federation's ethnic composition to the advantage of the Chinese; the mathematics of the merger suggests that the Malays welcomed Sarawak and Sabah to the Federation in order to counterbalance the demographic "Chineseness" of Singapore. Many people in Singapore opposed strongly the idea of "Malay Malaysia", with special privileges for indigenous peoples that excluded the Chinese and Indians, unsurprisingly given the proportions of ethnic groups in Singapore: 75% Chinese, 15% Malays, 7% Indians. The issue proved insolvable and after two years of attempts to find a compromise between indigenous and non-indigenous interests, Singapore[9] was expelled from the Federation.

Thus, since 1963, Malaysia has had two chief categories of citizens: the Bumiputera ("sons of the soil", or the natives) and the non-Bumiputera: Chinese and

Indians. However, it is important to underscore that the Bumiputera comprise two distinct groups: Muslims (mostly Malays, but also converts from animism, most notably the Melanau in Sarawak); and non-Muslims (chiefly the indigenous peoples of Sarawak and Sabah, who are predominantly Christian). Significantly, Malay is defined in the Malaysian constitution (art. 160) by three elements: as a person who professes Islam, habitually speaks the Malay language and follows Malay customs. Conversion out of Islam, in this light, equals ceasing to be Malay. Similarly, conversion to Islam, using the language and following the customs are conditions to be fulfilled by any Malaysian citizen who wishes to become Malay (and enjoy the constitutional reservations for the Bumiputera). The definition opens a way to become Malay; in the case of the disadvantageous proportion of Malays to non-Malays in particular regions, certain incentives can be introduced to attract desirable groups to become Malays.[10]

Sabah and Sarawak, much like Penang and Malacca in West Malaysia, have no Malay ruler. The corresponding function in the four states is carried out by their respective Yang di-Pertua Negeri, who are, however, not eligible for election as the federal Yang di-Pertuan Agong, or the king (one of the Malay rulers in the states). The Yang di-Pertua Negeri was officially called governor in the early years (and on occasion is still referred to as such), as the position corresponded to the British-established governors. The Yang di-Pertuan Agong appoints the Yang di-Pertua Negeri after consultation with the chief minister (art. 1. Constitution of Sarawak). There is no provision that the Yang di-Pertua Negeri has to be a Muslim; nevertheless, so far all the persons holding the office in Sarawak have been Muslims, but not all of them were Malay.

The indigenous peoples of Borneo were granted Bumiputera status, equalling them with Malays, but they were also denied language and education privileges that the Chinese and Indians in West Malaysia enjoy. At the time of the creation of the Federation of Malaysia, Sarawak "accepted Malay as the national language" (Leigh 1974, 89), but opted for a 10-year clause allowing use of the English language for official purposes. The Malay elites of West Malaysia pressured the Bornean states to introduce Malay as the official language and the language of instruction in schools much sooner, in order to boost national unity. However, Malay speakers were a minority in Sarawak and Sabah, while the non-Muslim indigenous spoke multiple languages/dialects. In particular the first Sarawak government took a strong stand towards the language and education policy. Given that "the Dayak literacy rate in English (1960) was three times their rate in Malay" (Leigh 1974, 89), it was believed to be more crucial for the natives to expand their literacy in English than to switch entirely to Malay (Leigh 1974, 93). A practical reason for this insistence on English education lay in the fact that at the time no schools in Sarawak taught in the Malay medium; Mandarin was the language of instruction in Chinese primary schools, while the other primary and secondary schools used English or some vernacular in the lower primary levels (Milne and Ratnam 1974, 46). These strong arguments fell on deaf ears in Kuala Lumpur; in 1974 the Sarawak legislature passed the bill that established Malay as the sole official language in the state.

Malaysia is a constitutional monarchy with a bicameral parliament. The power is vested in the lower house, called the Dewan Rakyat (the "People's House"). The Dewan Negara, or the Senate, has very limited powers[11] and all its members are nominated.[12] The king's position is politically symbolic, although he gives his assent to legislation (so do the rulers in their respective states). Malaysia is a federation of 13 states, or, as Milne suggests, a state with "federal *form*" (1967, 75, emphasis in original). Principally, the constitution can be amended by the federal parliament without obtaining consent from the state assemblies. The only exception (added upon Sarawak and Sabah joining the Federation), are certain matters referring to the Bornean states, in the case of which the respective assemblies must pass legislation consenting to the amendments. In the latter case the governor of the state must express his consent; he has to, however, listen to the state government's opinion in the matter (Milne 1967, 77). In general, "the federal government has more substantial powers by far than the states" (Milne 1967, 77). Moreover,

> after declaration of Emergency, the federal Parliament may make laws with respect to any matter on the state list [of powers], except matters of Muslim law or the custom of the Malays, or with respect to any matter of native law or custom in a Bornean state.
>
> (Milne 1967, 77)

The Bornean states were granted wider autonomy in some matters than other states (Milne 1967, 79). Most notably, the states have a veto on entry and residence of persons from other Malaysian states (whereas the control of immigration into the Federation remains under federal purview). No formal limitations of creating regional parties exist in Malaysia; to the contrary, it is the federal parties that need to register in the state if they wish to operate in Sarawak. Indeed, only in 1978 was the first non-Sarawakian party allowed to establish its branches in the state (Chin 1996c, 390). The Registrar of Societies, the agency responsible for party registration and, at the same time, the controlling institution for party activities, has often used its powers to either accelerate or stall the registration process according to the government's political needs (Chin 2002).

2.3 Political parties' origins

Prior to 1841 Sarawak was part of the north Bornean territory under the rule of the Sultan of Brunei. In 1841 a British cavalry officer, James Brooke, helped the Sultan crush a rebellion in the western part of the territory and, in return, was granted the title of Raja of Sarawak. His descendants governed the territory until 1946 when the last of the Brookes ceded Sarawak to the British government; Sarawak became a British colony (Chin 1996a, 16–17). In 1963 it joined Malaya, Sabah and Singapore and the four entities became the Federation of Malaysia. Prior to the creation of Malaysia, given the lack of structured organizations among the communal groups of Sarawak, except the Chinese, the matter of setting up political representation in the form of parties posed some difficulty, especially in terms of financing. As Roff noted, "all groups, the Muslim settlers and Chinese urban dwellers no less than the up-river groups deemed 'tribal', were led into modern political activity by members of traditional elites" (1974, 7). The financial support necessary to form political parties by those leaders came from the British in the form of timber licenses (Roff 1974, 8).

In this light we would be compelled to see the subsequent party establishment and the primary activation of identities as the result of two factors. Firstly, the already existing personae of power (Penghulus and Temenggongs, Sarawakian traditional leaders of ethno-regional components) were going to capitalize on their positions. Significantly, the traditional leadership was characterized by not only being limited to the co-ethnics (each linguistic/cultural community maintained its own hierarchy of power), but also to a particular region in which the leaders operated, usually a river basin. The second factor is, naturally, the financial capital and its availability in Sarawak at the time from only two sources: timber licenses and Chinese businessmen. The nascent party organizations were therefore bound to reflect these two factors: traditional leaders with their communal backing, and modern political entrepreneurs dependent on the financial backing of Chinese business elites.

In the years that preceded the creation of the Federation of Malaysia, the Sarawakian parties were primarily split along the idea of joining Malaysia. In 1963 the dividing factor was Sarawak's autonomy and position within the Federation – this was the chief party cleavage of the 1963 Sarawak District Council elections (Milne and Ratnam 1972, 83), and this cleavage cut across ethnic, linguistic, religious and regional affiliations. None of the then-established parties was exclusively oriented towards one particular ethnic identity; either "multi-racialism" was envisaged as a mobilization strategy, or the parties sought close cooperation with another ethnic category with which they believed they had common interests.

The question of power relations between the communal categories was nevertheless open. At that time, the Sarawakian ethnic split was roughly 24% Muslim, 44% non-Muslim indigenous and 30% Chinese. The Chinese in Sarawak were the only group prior to 1963 that had established functioning organizations that comprised and penetrated significant portions of the community (Leigh 1970, 191). The impetus to create the Sarawak United People's Party (SUPP) – the first organization of this sort in Sarawak – came from the British government, which intended it to be multi-racial. Ong Kee Hui and Stephen Yong, Chinese from Sarawak's capital of Kuching, picked up on the idea and despite their Chinese background they intended to spread the party's activities across the ethnic divisions (Leigh 1974, 13–14). As a result the Sarawak United People's Party was established in 1959, although by then the British government had lost its original enthusiasm because of the communists' interest in the project. SUPP found it easier to recruit Chinese members, in particular those from the Kuching area, although it enjoyed significant non-Muslim native support and negligible Muslim support. Most of SUPP's elected representatives in 1970 were Hakka and Hokkien; other party executives were Teochew and Cantonese (Leigh 1974, 20–21). Foochows were absent among the party's heavyweights.

However, as with many parties in Sarawak, throughout the decades of its operation SUPP was at times an ethnic, at times a multiethnic and at yet other periods a non-ethnic party. Until SUPP joined the Alliance (the coalition of parties ruling in Malaysia, the predecessor of Barisan Nasional) in 1970, its image was constructed of three elements: "anti-Malaysian" "leftist" (communist-penetrated), "Chinese" and "Hakka/Hokkien" (in this order), although the party made conscious efforts to attract non-Muslim indigenous and maintain its multi-racial image. Needless to say, the "leftist" and the "Chinese" in some instances went well together: when the Malaysian proposal was announced in 1962, SUPP not only upheld that an independent state would be more suitable for the Bornean states in general, but also criticized the specific issues of the Malay constitution that would impact Sarawak and Sabah the most: "the inclusion of constitutional amendments which would discriminate against the non-indigenous, that is, the Chinese [. . .] Islam and Malay are unacceptable as the national religion and language, respectively" (Chin 1996a, 66). While the contention of native privileges was clearly a reflection of Chinese interests, the protests against language and religion were equally crucial to non-Muslim natives and cannot be read as specifically Chinese.

The Hakka/Hokkien (and to a lesser extent Cantonese, Teochew) inclination of SUPP is discernible primarily in the composition of the party's chairmanship and its elected candidates. Kuching-based Chinese were strongly overrepresented among the party's executive, which is informative in the light of the fact that there are almost no Foochow settlements in the First Division and in Kuching itself (Leigh 1974, 58–59); non-coincidentally, Foochows were excluded from the party's leadership. There is, however, little evidence to support the Hakka/Hokkien ethnic character of the party based on election results of that period: in 1963 SUPP polled well both in the First and in the Third Division, despite vast Foochow settlements in the Third Division. The dialectical cleavage, however, should not be understated. SUPP between 1960 and 1970 was a multi-dimensional party, with a clear ideology and a programme that extended beyond simple ethnic particularism; however, the party's later evolution points to the viability of the dialectical division.

Chin, who generally maintains that SUPP from its inception was a Chinese ethnic party (1996a, 81), also admits that "among the 205 candidates fielded by SUPP in 1963 there was a mixture of mainly Chinese candidates in the Chinese urban areas and indigenous candidates in mixed and rural areas" (Chin 1996a, 73). Note that the "skillful government campaign to discredit SUPP as a Chinese and communist party meant that more of the indigenous members were persuaded to leave the party" (Chin 1996a, 81–81). Milne and Ratnam show that in terms of ethnic membership, SUPP was attracting more new members from among the native groups than the Chinese (1974, 75). Therefore, during the 1960s, SUPP was an opposition party, popular among all Chinese dialectical categories except for Foochow and, albeit less so, non-Muslim indigenous. The party also had leftist inclinations and was perceived as "Chinese".

The Sarawak Chinese Association (SCA) was established in 1962 with the help of the Malaysian Chinese Association (MCA), the Alliance component with

Chinese background in West Malaysia. Members of SCA were large-scale businessmen and the party was dominated by Foochow-speaking Chinese (Leigh 1974, 204). SCA's explicit message and name qualifies it as a Chinese ethnic party; however, its implicit message and the quarters to which it appeals make it a "Foochow Chinese" party. "By 1967, well over half of the [SCA] Committee members lived in the Third Division and just over half of the twenty-three Central Committee members were Foochow. It seems that SCA is in fact striving to stimulate support based upon dialect group, and to establish itself as the party for the rich young aspiring Chinese executive" (Leigh 1970, 205–206). SCA's appeal to the Foochows coincided with the category's lack of representation through SUPP.

Because of SUPP's strong anti-Malaysia and anti-Alliance stand prior to 1970, the SCA was safe as a coalition member representing the Chinese strand of the society in the government; despite the SCA's weak performance in the 1963 election and its low popularity in the society, SCA was given ministerial portfolios in the first cabinet. Therefore, while appealing to the Foochow component of the Chinese quarters in Sarawak and enjoying support from a geographically limited area, within the government the party was not a "Foochow Third Division" party, but a "Chinese" party. Therefore, it is safe to assume that both SUPP and SCA within the Chinese community were associated with respective dialectical identities, while for the non-Chinese, they were simply "Chinese parties".

The first parties representing the Muslim indigenous also reflect the ethnic complexity within that category. In West Malaysia the "Malay", "Muslim" and "Bumiputera" categories are coterminous. Not so in Sarawak; "Muslim" comprises Malays and most, but not all, Melanau, Kedayan and Bisaya, while "Bumiputera" includes all of these and all other non-Muslim natives, who actually outnumber the Muslims. For the Muslim community in Sarawak joining the state of and for Malays, i.e. Malaysia, meant that their position became incredibly complicated. The Sarawakian Malays became Malaysians just when Malaysia ceased to be simply a Malay state, and – at least according to one interpretation – it became a "Bumiputera state". In Sarawak, Malays did not enjoy the same titular nation position as the Peninsular Malays did. Here they had to share the spoils with non-Muslim indigenous, and the question was open as to the position of the non-Malay Muslims. Another source of apprehension was related to the possibility of the West Malaysian Malays, visibly very well organized and strong, overpowering the local elites and stripping them of power.

Around the time when Sarawak joined Malaysia there were two strains of elites among the Muslim community. One strain comprised the traditional aristocratic leaders within the Kuching (First Division) area, "government recognized, of reasonable means, and frequently appointed by the government to representative institutions" (Leigh 1974, 27), who were committed to the idea of Sarawakian Sarawak. These leaders, most notably Dato Bandar, established the Parti Negara Sarawak (Party of the Sarawak State, or PANAS), with the will to attract support from across the entire ethnic spectrum. The second strain of Muslim leaders included young professionals, some foreign-trained, mostly from the Third Division and many of Melanau extraction, who sought quick political gains. These leaders established Barisan Ra'ayat Jati Sarawak (BARJASA): a party with Muslim leaders championing the interests of all "Sarawak races", but not counting the Chinese among them (Leigh 1974, 31).

The geographical division of the two parties is particularly clear if we compare the parties' performance in the 1963 election; in the First (predominantly Malay) Division, there were 43 councillors from PANAS and only 9 from BARJASA. In the Third Division, with a high proportion of Melanaus, PANAS won only one councillor's seat, against BARJASA's 26 (in all cases the numbers include independent candidates supported by the respective parties) (Leigh 1974, 71). PANAS, whose leadership was attracted both to the Malay/Islamic outlook of Malaysia, and to the idea promoted by the Sarawak National Party (SNAP, see later) of a "stronger Sarawakian stance against the efforts of local 'upstarts' [i.e. BARJASA] who had the support of Kuala Lumpur" (Leigh 1974, 98). Clearly, over the years the anti-Malaysia stance would become less and less relevant,[13] but the cleavages that coincided with the pro- and anti-Malaysian attitudes were not easy to overcome. Traditional Muslim elites in the state were apprehensive of the young conformist Malays and Melanaus who quickly found their way to the top of state and federal governments with the unofficial backing of the United Malays National Organization (UMNO), the Malay component of the coalition ruling in West Malaysia. In other words, the loyalty among Muslims was split between peninsula-sponsored affiliations implicitly based on religion on one hand, and pro-Sarawak, multi-racial solutions on the other. PANAS' existence in those early years as a Muslim party reflects one of the many divisions in Sarawak that extend beyond simple ethnic understanding; however, it is important to underscore that the division was not between the Malays and Melanaus. Abdul Taib and his uncle Abdul Rahman Yakub, both Muslim Melanaus, were the two most prominent leaders of BARJASA and had an important choice to make. As the leaders of their party they had to pick an identity from the categories available to them to gain as loyal a support as possible and at the same time not estrange any vital pockets of the electorate that were in the party's reach. They dismissed the "Melanau" category as too narrow a support base[14] (Milne and Ratnam 1974, 91; Roff 1974, 29) and chose the "Malay" identity for the party, keeping, however, the door open for non-Muslim natives, thus playing with the "Bumiputera" concept.

Around the time of the first election in 1963, among the non-Muslim natives, the Ibans were the only ones who became politically active to the extent of forming their own parties. Witnessing the emergence of SUPP with its strong Chinese support, and BARJASA's and PANAS' Muslim background, a group of Ibans – employees of the Shell company – established the Sarawak National Party (SNAP). SNAP, much like SUPP and PANAS, had strong ambitions to be a multi-racial party, or at least to be a non-exclusive one. However, seeing that the competition was displaying clear communal orientation, SNAP quickly realized that the Ibans were its niche in the market. So was the leadership, concentrated around the persona of Stephen Kalong Ningkan.[15]

Ningkan was an Iban from the Second Division, from the Saribas river basin. SNAP soon acquired an image of a party that champions the interest of the Second

Division Saribas Ibans, who were mostly Anglicans. Since the Brookes had allocated different regions of Sarawak to different Christian missions, Christian denominations coincide with regional borders, and Anglicans are mostly found in the Second Division. In this area, as Leigh points out, the primary division (if not animosity) was between Malays and Ibans, while the Chinese were relatively few and there were few grounds on which Ibans and Chinese could compete (1974, 83). Therefore, SNAP accepted Chinese as members and was indeed financed chiefly from Chinese sources (Searle 1983, 20). For these reasons, and because of upcoming political developments, SNAP found it easier to cooperate with the Chinese, both within and outside of the party, than with Malays. SNAP from its inception until the 1966 crisis was therefore an ethnic party that appealed to the Second Division Anglican Saribas Ibans and, strategically, to the Chinese.

Unlike the Second Division Ibans, the Third Division Ibans of the Rajang river are mostly Methodists and Roman Catholics. In the Third Division, unlike in the Second, the Ibans did not see eye to eye with local Chinese, who competed with the Iban for land. The most influential of the Rajang Ibans was Temenggong Jugah, at the time the highest-ranking of the indigenous leaders in Sarawak. Although Jugah had already joined PANAS, upon an initiative from his fellow Rajang Ibans, he welcomed the idea of a new party and became its president; this way in 1972 Party Pesaka (Parti Pesaka Anak Sarawak or Traditional Sarawak's Children Party) was established (Leigh 1974, 36). Milne and Ratnam pointed out that "which part of the country one is from is often as important as what community one belongs to, as testified by Pesaka's emergence as a party which sought to represent Third Division Ibans" (1972, 145).

Indeed, although Ibans speak the same language and share several other cultural markers, the "Iban" comprise at least two identities, "Rajang" and "Saribas", and it can hardly be overestimated that these two categories were very much activated in the 1960s. Pesaka originally accepted only non-Muslim indigenous, but in 1965 also welcomed Malays as members, but never Chinese (Leigh 1974, 84). Therefore, I maintain (in accordance with Leigh (1974) and Searle (1983)) that at that time SNAP and Pesaka were respectively Saribas Iban and Rajang Iban ethnic parties, based on their agreement "to limit recruiting on the other's geographical base" (Leigh 1963, 45). Both parties were supported chiefly by Ibans, i.e. not by Malays or Chinese, in the 1963 election (85% and 91% of their respective support came from the Ibans).

However, in its explicit message Pesaka appealed to "Dayaks", not "Ibans" or "Rajang Ibans". The objectives of the party were:

> To assist all Dayaks to unite in pursuing the common aim and interest with the object of promoting and presenting an [*sic*] unifying approach to problems which affect their people in the successful government of this country.

> To preserve the heritage of Sarawak having regard to the necessity of promoting the political, social, economical and cultural advancement of the Dayaks in a constantly changing world.

> To ensure by all constitutional means the Dayaks have a rightful say in the government of the territory.
>
> (Leigh 1974, 37)

Based on the explicit message the party conveyed, we would have to conclude that Pesaka is a "Dayak" ethnic party. I will contend, however, that despite this clear message, the purported "Dayak" image of the party has to be seen in the wider perspective of two facts. For one, in order to be a serious contender on the Sarawakian political scene the party would have to expand its base beyond the Rajang Ibans. The only unclaimed wider category was "Dayak". Secondly, the party was arguably positioning itself against SNAP, which on the face of it was campaigning for all Sarawakians, but clearly was a Saribas Iban party. To officially appeal to all "Ibans" would be unwise of Pesaka, as it would unnecessarily antagonize SNAP, and so "Dayak" became the official line of the party. However, I argue, "Dayak" did not become a politically activated category until the early 1980s, while Pesaka, judging by its leadership and area of political contest, was a Rajang Iban ethnic party.

The first activated ethnic categories in Sarawak prior to joining Malaysia comprised the following: non-Foochow Chinese; Foochow Chinese; Malay/Melanau; Rajang Iban and Saribas Iban. These categories were activated by political parties, but in the parties' agendas the particularistic ethnic interests were combined with programmatic strategies (e.g. stance towards Malaysia) as well as with visions of inter-ethnic relations. These categories were activated before institutional incentives were known to the players, and some parties (e.g. SCA) were already gearing up for competition within a consociational political solution, while others (SUPP, SNAP) envisioned multiethnic parties competing against each other. Two parties, PANAS and BARJASA, appealed to both Malays and Melanaus, but, as the first election showed (see later), PANAS was more popular among Malays, while BARJASA was more popular among Melanaus. The ethnic background of the parties' leadership was decisive when it came to establishing the parties' geographical reach, which in return resulted in the parties' popularity. Significant pockets of the society had not been mobilized within the existing party structure; the 30% of votes cast for independents in the 1963 election suggests that the Fourth and Fifth Divisions were up for grabs; up until this point, no party had claimed the non-Muslim indigenous voters who were not Iban.

2.4 The first election and the first government

The first relevant election in Sarawak based on general franchise took place just before the territory joined Malaysia in 1963. The polling was carried out according to "the three-tier system, by which district councils and, in turn, divisional advisory councils acted as electoral colleges"[16] (Leigh 1974, 49). On the top of the electoral pyramid (the third tier) members of Council Negri[17] (or State Council, i.e. the legislative assembly) were elected. The 1963 election to the Council Negri was therefore indirect; voters directly elected only the district councillors in 24 districts, who in turn elected divisional councillors in 5 divisional councils, who then elected 36 state councillors.[18]

Eligible to vote were all persons 21 years of age or older who resided for at least 7 out of the previous 10 years in Sarawak (regardless of their citizenship (Leigh 1974, 51)). "The actual constituencies were district council wards, characteristically single-member, though in the towns there were some multi-member constituencies. The overall district council boundaries were coterminous with the limits of administrative districts. The only exceptions to this rule were that two councils were formed in each of the following districts: Kuching, Sibu, Miri and Binatang" (Leigh 1974, 51). The margin left for gerrymandering was therefore minimal and limited to the "representation accorded to each district council at the next level of administration" (Leigh 1974, 51). The 24 members of the national parliament were selected by the Council Negri, although not necessarily from among its members. MPs were nominated according to the number of seats each party held in the state assembly.[19]

Before Malaya, Sarawak, Sabah and Singapore became the Federation in 1963, Malaya had already established its political *modus operandi* in the form of a multiethnic coalition, at the time under the name "the Alliance". Those of the Sarawakian parties that were in favour of the idea of Malaysia found it of advantage to create a local coalition corresponding to the one in the peninsula. PANAS, BAR-JASA, SNAP, Pesaka and SCA all joined the Sarawakian Alliance in 1962. Just prior to the 1963 election, PANAS left the Alliance on the state level because of a dispute over the Alliance's state leader position. Parties within the Alliance were not to compete against each other; therefore, the main competition was between the Chinese parties and the Malay/Melanau parties, as one of both was outside the Alliance. However, SUPP also fielded candidates in non-Muslim indigenous areas, and so did PANAS, as both were trying to expand their bases to these quarters and were allowed to do so, not being limited by coalition restrictions.

Upon completing of election procedures in district and divisional councils, the final composition of the Council Negri (the state legislature) was SUPP – 5 councillors, PANAS – 5, BARJASA – 6, SNAP – 7, Pesaka – 11, SCA – 3, Independents – 2 (computed based on Leigh 1974, 78, table 20). The proportion of votes each party obtained is shown in Table 2.1.

Table 2.1 1963 Sarawak election results by division and party

Division	Votes received by party				Ethnic category of registered voters		
	SUPP	*PANAS*	*Alliance*	*Independent*	*Chinese*	*Malay/Melanau*	*Dayak*
First	31.7%	32.8%	22.9%	13.0%	34.9%	31.9%	33.1%
Second	10.6%	17.2%	52.0%	20.1%	8.6%	24.3%	67.1%
Third	23.0%	1.2%	39.0%	36.8%	26.3%	17.4%	56.2%
Fourth	14.2%	10.5%	22.4%	53.0%	20.3%	23.1%	56.5%
Fifth	0.0%	0.0%	23.2%	76.7%	10.0%	50.7%	39.2%
Total	21.4%	14.3%	34.2%	30.2%	24.0%	24.9%	51.1%

Source: Leigh (1974, 57).

The results of the first election in terms of ethnic performance by party are difficult to analyze because of several facts. Firstly, only on the first, lowest tier (ward) were the elections direct. The members of the divisional councils and the Council Negri were elected indirectly. Secondly, almost a third of all elected representatives were independents, who only after the election were courted to join parties in the Council. Naturally, the ruling parties found it easier to co-opt the independents. Candidates who won unopposed filled a sixth of all seats. However, Leigh found that SUPP was very strongly positively correlated with Chinese votes and Malay constituencies voted strongly for PANAS, while Melanaus voted for the Alliance (and BARJASA within it) (1974, 61–63). Most votes for the Alliance came from the Iban quarters. The Rajang Ibans delivered the most votes to the Alliance in their area through Pesaka, and the Saribas Ibans voted for SNAP within the Alliance. In the first election, no party appealed particularly to the Land Dayaks (later known as Bidayuh) in the First and Second Division, but SNAP and BARJASA were the two most popular parties in these Bidayuh-majority areas (Leigh 1974, 64).

The very first resolution the Council Negri passed when it convened in September 1963 was to join the Federation of Malaysia. Five members of the Council, all from SUPP, were against the decision. Sarawak officially became part of the Federation on September 16, 1963. The bodies elected in 1963 served until 1970 (two years longer than their constitutional term), for two reasons. First, Sarawak was to hold its elections simultaneously with the entire Federation, and these were due only in 1969. Second, when the elections did take place in 1969, before the polling was conducted in Sarawak, a state of emergency was declared because of riots in Kuala Lumpur[20] and the electoral process was completed only in 1970.

The constitution of Malaysia regulates the matters of substantial powers of the federal and state governments, but the actual *modus operandi* of state–federal relations for non-policy matters was to be established through practice. Especially the level of political autonomy of the Bornean states had to be cleared. The first test of state–federal relations occurred when the positions of the first governor and chief minister were to be decided. The Alliance parties won the election, but neither of them could claim a victory decisive enough to have an unquestionable claim to the chief ministership. Given the ethnic composition of the state, "the only real question was which Dayak would be the Chief Minister" (Leigh 1974, 79). The parties constituting the ruling coalition in the state at the time agreed to nominate the leader of one of the Iban parties, Stephen Kalong Ningkan, the chief minister, and the leader of the other Iban party, Temenggong Jugah, as the governor. Temenggong is the title of the highest-ranking Iban leader, and, in order to pay respects to local tradition and legitimize the government, Jugah was considered indispensable as a member of the executive.

However, according to the London Agreement, which stipulated the conditions of the merger of Sarawak and Malaya (and Sabah and Singapore), the Malayan Yang di-Pertuan Agong and the British queen would jointly nominate the governor of Sarawak. The directive from Kuala Lumpur was clear: if a non-Muslim were to be the chief minister, a Malay would have to be nominated for the governor position. This way, however, Temenggong Jugah would be left without any significant post.

The nomination was scheduled to happen just a few days before the official establishment of the Federation of Malaysia was to be announced and Jugah, whose Pesaka party was an important member of the state-level Alliance, almost decided to withdraw his party's support for the Malaysia plan (Leigh 1974, 79). The solution to this impasse was the creation of the position of Minister for Sarawak Affairs in the federal cabinet, which Temenggong Jugah was appointed to. Later on, it was established via practice that the head of the Alliance (and later its successor, Barisan Nasional, or the National Front) has "the power to choose Alliance chief ministers"[21]; how this general rule applies to contemporary Sarawak will be shown in this chapter.

Stephen Kalong Ningkan's 1963–1966 cabinet was definitely an Iban-led one, and Saribas Iban at that. The full composition of that suspenseful Sarawak state cabinet is well worth consideration:

> Stephen Kalong Ningkan (SNAP), Chief Minister, Second Division Iban
> James Wong Kim Min (SNAP), Fifth Division Hakka Chinese
> Abdul Taib bin Mahmud (BARJASA), Third[22] Division Melanau
> Dunstan Endawie anak Enchana (SNAP), Second Division Iban
> Teo Kui Seng (SCA), First Division Teochew Chinese
> Awang Hipni (BARJASA), Third Division Melanau
>
> (Leigh 1974, 82)

The composition of the first government could also be put in these terms: "2 Dayak, 2 Chinese and 2 Muslims".[23] This was duly noted by Leigh (1974, 82), i.e. "Dayak", "Muslim" and "Chinese" were valid terms to be used in discussion of distribution of political power, but arguably the geographical division of the Ibans also played a role; so did the linguistic background of the Chinese. The composition of the cabinet itself is more an outcome of programmatic and/or personal reasons than of ethnicity itself. With Ningkan as chief minister, "the government came to be primarily British-influenced SNAP and SCA combination of Dayaks and Chinese. [. . .] The Muslims, the third division Iban [i.e. Pesaka], and their federal supporters were far from the corridors of state power" (Leigh 1974, 83). The Muslim voice in the cabinet (represented by two Melanaus) was mostly circumvented and the PANAS Malays were absent from the cabinet. Non-Muslim indigenous, other than Second Division Iban, were also absent. Therefore, although some multiethnic decorum was maintained, resulting from the Alliance coalition arrangement, Ningkan did not seem to be bothered too much to appoint token Malay or Rajang Iban ministers (let alone other non-Muslim indigenous).

In the early years, the decision of the prime minister about whom to nominate to his federal cabinet did not always reflect the power relations in the state, whether expressed in terms of ethnicity or political parties. Between 1963 and 1966 the policy-making core of the Sarawakian cabinet consisted of members of two ethnic categories, Saribas Iban and Chinese (and British expatriate advisors);

however, the ministers appointed to the federal cabinet at the time were of Mela-nau and Rajang Iban background (Leigh 1974, 83), which were precisely the two categories side-lined in the state cabinet. This discrepancy was not random or coincidental; the appointments to the federal governments were at the discretion of the prime minister, whose ethnic and party preferences, when it comes to fed-eral cabinet composition, were different from voters' preferences that resulted in the particular electoral outcome in Sarawak in 1963. The consociational modus had yet not settled in Sarawak and exclusive cabinets were perfectly conceivable.

In 1965 Taib Mahmud staged a mini coup which, if successful, would have removed Stephen Kalong Ningkan from his position as chief minister. The imme-diate opportunity to make a move to oust Ningkan was provided by a bill tabled by Ningkan's cabinet minister from SCA. The bill pertained to the ever-sensitive and most crucial issue of land ownership in the state, an issue that was and remains a thorn in Chinese–indigenous relations in Sarawak. The Chinese have very lim-ited access to the land, which they cannot buy, the land being reserved to the indigenous residents[24] since British times. The bill, which would have allowed the Chinese to buy land, was withdrawn, but Taib Mahmud sensed the opportunity of turning the issue against Ningkan. The BARJASA leader proposed establishing an alternative coalition, consisting of native parties only (these were BARJASA, Pesaka and PANAS), from which both SNAP (because of its multi-racial; i.e. Chinese-inclusive, membership) and SCA (as a Chinese party) would be excluded. Such a coalition would be a realization of the "Bumiputera" concept, which, in the end, was introduced specifically as a platform corresponding to "Malay" in West Malaysia. Such a "Native Alliance", as Taib Mahmud named it, would be parallel to UMNO in West Malaysia.

The "Native Alliance" did not take off and Ningkan prevailed, and this was mostly thanks to the support of Pesaka's Ibans and Temenggong Jugah (the Sarawakian Alliance's chairman at the time, as well as the federal minister of Sarawak affairs) in particular. Jugah is quoted as believing that Taib's intention was to create "a split between Dayaks and between the races" (Milne and Ratnam 1974, 222). Pesaka's gain from the situation was to have two of its members appointed to the state cabinet. Taib Mahmud was temporarily dismissed as state minister for plotting against the chief minister, but Ningkan had to reinstate him supposedly because of strong advice from Kuala Lumpur (Milne and Ratnam 1974, 222). When BARJASA finally succeeded in ousting Ningkan in 1966, his party, SNAP, moved to the opposition, along with all the Saribas Ibans.

2.5 Critical juncture I: state-federal elections

The expulsion of Singapore from the Federation was an alarming signal to Sarawak and Sabah. The idea of "Malaysian Malaysia" promoted by the People's Action Party (PAP), the strongest Singaporean party, did not sit well with the Alli-ance government in Kuala Lumpur, and after several months of propaganda war the Malaysian prime minister decided that the differences between him and Singa-pore Chief Minister Lee Kwan Yeu were irreconcilable. Singapore was expelled from the Federation in August 1965,[25] which led the Sarawak United People's Party, the main opposition party in Sarawak and the strongest opponent to the idea of Malaysia, to ask: "What will be the position of the Borneo states if there should emerge governments there not so pliable to Alliance ways?" (Ongkili 1985, 187).

This question was to be answered very soon. The second test of the limits to federal influences in Sarawak occurred in 1966. The intra-cabinet tensions had been growing for a while and the Muslim indigenous members were not seeing eye-to-eye with Iban Chief Minister Stephen Kalong Ningkan. The Alliance government in Kuala Lumpur also found it difficult to cooperate with Ningkan; he was an uncompromising partner in federal policy implementation and the afore-mentioned education and language policy met especially strong opposition from the state cabinet and the chief minister. Therefore, in June 1966 the prime minister decided to entertain the request from 21 rebellious members of the Sarawakian Council Negri to back their plan to oust Ningkan. The outcome that the Alliance was hoping for was for the chief minister to resign. But Ningkan refused to step down and instead requested that the governor dissolve the assembly and declare elections; the request was declined because the new constituencies for direct elections had not yet been delineated. The governor, upon advice from the national Alliance leaders, dismissed the cabinet.

This was followed by an impasse more than three months long, during which the newly appointed Chief Minister Tawi Sli (also Saribas Iban but from the Pesaka party, which was much more conciliatory towards Kuala Lumpur than Ningkan's party) could not take up his duties while the old chief minister refused to leave his position claiming his dismissal to be unconstitutional. To resolve the impasse, a state of emergency in Sarawak was declared on September 15, 1966, which allowed the federal parliament to enact two amendments to the Sarawakian constitution. One gave "the Governor the power to convene the Council Negeri at his own discretion" (Ross-Larson 1976, 52) (otherwise, the governor could call a sitting of the assembly only at the request of the chief minister). The other amendment empowered the governor "to dismiss the Chief Minister on a direct appeal from Council Negeri members" (Ross-Larson 1976, 52) (which otherwise would have required a vote of no confidence in the assembly). The Council convened and dismissed Ningkan, installing Tawi Sli as the new chief minister.

This incident represents the first critical juncture in the chronology of Sarawakian politics. From this point on it was clear that it is within the ruling coalition that state–federal relations are decided, and not within the constitutional prescriptions in the matter. The fact that Sarawak has its own set of parties that constitute the coalition is only limitedly relevant. Therefore, the Alliance, and later the Barisan Nasional, has to be seen as a super-structure with pseudo-institutional powers. While the ruling coalition is not an institution in the constitutional sense, it does provide the central government with a direct channel to induce courses of action at the state level that are convenient for Kuala Lumpur. In this sense the ruling coalition is not only, as usually seen, an instrument of co-opting ethnic components and maintaining the power-sharing scheme, but also an instrument of strengthening the control of the federal government over the states. As Milne put it, "The best guarantee of happy federal-state relations does not lie in any constitutional provisions but rather in the harmonizing [*sic!*] influence of membership of the same party" (1967, 79).[26]

The Alliance/Barisan Nasional provides a meta-level of politics in Malaysia as a quasi-institution and supra-party. To bring the argument home, the position of the prime minister should be understood. Under the rule of the Alliance/Barisan Nasional, it is the UMNO president who becomes the prime minister of Malaysia. Therefore, as long as there is no UMNO in Sarawak (and the Sarawakian elites have little interest in inviting UMNO to the state), there is no practical possibility of a Sarawakian becoming the prime minister. Similarly, because important nominations for positions in Sarawak lie in the hands of the prime minister/ruling coalition president and the Yang-di Pertuan Agong, who are both Muslims (the first by the principle of UMNO, the latter by the rule of the constitutional provision), it is unlikely that under the current coalition reign Sarawak would have a non-Muslim chief minister[27] or a non-Muslim Yang di-Pertua Negeri. These, albeit informal, rules will be shown to offer powerful explanations of the development of ethnic power relations in Sarawak.

State–federal relations are part of the institutional set-up and as such an exogenous variable. However, the facts listed earlier suggest that it is the party system in the form of a permanent coalition that became the tool for maintaining state–federal relations. The Alliance/Barisan Nasional, more than a coalition, is a *modus operandi*, a mechanism for managing the power relations between the centre and the federal states; a mechanism more effective than the constitutional regulations in the matter. The 1966 intervention showed that the central leadership was willing to go to great lengths to ensure smooth cooperation between Kuala Lumpur and Kuching, and the already existing Alliance loyalties played an important role in the process. As Milne pointed out, "the removal of Ningkan [. . .] was achieved by working through the Malaysian Alliance machinery, not the Sarawak Alliance machinery" (1967, 86). After the Ningkan cabinet crisis of 1966, political pragmatism instructed that were the state government to be unstable, weak, vulnerable or arrogant and uncooperative, the federal Alliance/Barisan Nasional would find a way to replace it with a more compliant one.

After Ningkan was toppled, Pesaka members took over the slots assigned to non-Muslim indigenous in the cabinet. Tawi Sli[28] of this party became the chief minister, as the coalition found it important to avoid the impression that Ningkan's removal from office was an ethnic matter. Officially, at the top posts of the Sarawak Alliance were two Ibans, Temenggong Jugah as its chairman and Sidi Munan as the head of the Alliance secretariat in Kuching (*Far Eastern Economic Review* 1970).

The composition of the new cabinet was much more diversified in terms of indigenous membership, but included no Chinese members (Leigh 1974, 106):

Tawi Sli (PESAKA), Chief Minister, Second Division Iban
Abdul Taib Mahmud (BARJASA), Third Division Melanau
Francis Umpau (PESAKA), Third Division Iban

Abang Haji Abdulrahim (PANAS), First Division Malay
Awang Hipni (BARJASA), Third Division Melanau
Tajang Laing (PESAKA), Third Division Kayan

This would have been the single all-native cabinet in Sarawak's history, as neither BARJASA nor Pesaka saw much need to include Chinese in their cabinet. However, two months after the original line-up was announced, two SCA Chinese members were added to the cabinet, presumably to avoid possible adverse actions from Chinese quarters (compare Leigh 1974, 106–107). In the end, the cabinet was a continuation of the previous one in the sense of having an Iban as a chief minister, but the much stronger influence of the Muslim component was obvious.

Around this time all parties were learning to play by the new rules, which resulted from joining Malaysia and in particular being part of the national-level Alliance. The parties were also starting to gear up for the upcoming showdown in elections according to the West Malaysian electoral structure. In 1965, the BARJASA party leader at the time and later Chief Minister Abdul Taib said:

Perhaps we can take an example from the success on the mainland where unity begins with unification on a uni-racial basis first and then co-operation and understanding is forged between these groups of disciplined uni-racial organizations. [. . .]

BARJASA believed in unity among the natives first and having achieved that unity would then closely co-operate with non-natives [. . .] it is easier to co-operate with the non-natives when the natives are themselves united.

(quoted in Leigh 1974, 97).

BARJASA indeed believed in its chances to gain strong non-Muslim indigenous support; the Muslim leaders of the party "regarded their estrangement from the Dayaks as the product of British 'divide and rule' policies" (Leigh 1974, 84, fn. 6). The idea of merging the natives (which was later framed in the concept of "Bumiputera") seemed extremely tempting to BARJASA as this would be an easy way to shape Sarawakian politics according to the West Malaysian example. Side-lining the Chinese seemed the task of the day for the Muslims. In this light in the 1960s, BARJASA deemed it a good idea to establish UMNO in Sarawak and in this way unite all the natives within one strong party which was well-connected to Kuala Lumpur. "Tentatively the party was to be renamed United Malaysian Native Organization, to incorporate all native peoples within the party. The formation of a Sarawak branch would have a dual purpose, to unite all Muslims in one Sarawak organization, and to exert Muslim leadership over 'all we natives'" (Leigh 1974, 84). UMNO itself later discarded these plans, but they show the potential, which never really materialized,[29] to organize explicitly the natives or Bumiputera as an activated category.

In 1966 the two Malay/Melanau parties merged to form the Bumiputera party. The negotiations had been under way for several years. The understanding that the

Muslim categories could only win by being organized in one party was shared by the leadership, but there was little agreement between the parties about the leadership of the merged organization. The death of one of the PANAS leaders, Dato Bandar, and the move of the other leader, Abang Othman, to SNAP accelerated the merger, as at least the leadership question had been solved. As it happened, by the time the Bumiputera party came to life, the PANAS leadership seemed to have surrendered to BARJASA in the merger, although strong efforts were made to make it look like they are equal partners. The leadership of the Bumiputera party remained entirely Melanau/BARJASA (Milne and Ratnam 1974, 85–92).

At the same time, within one party, the solid "Malay/Melanau" category was activated, rendering irrelevant the competition between traditional and modern leadership. Nevertheless, the PANAS-BARJASA episode proves two important points, hardly visible beyond this point in Sarawak's history: there is no intrinsic disposition of Muslims to being more united than other ethnic categories in Sarawak, despite the popular understanding that their common religion brings them together. Even more so, the "Malay/Melanau" category fits mostly the main leadership of the Bumiputera party, who chose this category to identify with and to market among the Malays and Melanaus.

The period between 1963 and 1969 in Sarawak is critical for this study for three reasons. Chiefly, the first set of ethnic categories was activated at that time; the Malay/Melanau category was activated based on religious commonality, despite the regional distinction existing within the category. Within the Chinese two categories were activated: one was Hakka and Hokkien dominated and Kuching oriented; the other was focused on the Sibu-region Foochow-dialect category members. Among the non-Muslim indigenous, only the Iban speakers were politically organized. Despite the linguistic unity, they were split regionally into Rajang (river) Ibans and Saribas (river) Ibans.

Secondly, the polity was not a consociational one yet; parties were becoming increasingly ethnic-exclusive (BARJASA and Pesaka were ethnic from the beginning), but the ethnic particularism in their agendas was mixed with other cleavages (traditional vs. modern leadership, right-wing particularism vs. left-wing universalism). The existing ruling coalition was wobbly and parties within it were undermining each other, proving that the competition between them still existed and their spheres of influence were not yet defined, let alone fixed, as would be the case in a consociational polity. The "Bumiputera" and "Dayak" categories were showing their potential and parties were looking for ways to mobilize them to their advantage. In terms of identity shifts, Malays and Melanaus could choose to identify with either "Malay/Melanau" category (all Muslim), or opt for "Bumiputera" (which would also include non-Muslim indigenous). Ibans could identify according to their region (Saribas or Rajang), or as Ibans, or as the wider "Dayak", or also opt for "Bumiputera". The Chinese could choose to identify with their dialectical categories. Therefore, each Sarawakian saw several categories with which he or she could identify, and many political channels were open for activating any of these categories.

Lastly but not least, during the 1966 crisis the federal-level Alliance intervened to unseat the chief minister, who maintained his own vision of Sarawak within

Malaysia and the way its ethnic diversity is reflected in politics. The character of the ruling coalition as a supra-institutional entity was defined from then on. Obedience of the Sarawakian state government was coerced and an important precedence was established.

2.6 1969–1970 elections

Since Sarawak became a part of Malaysia, Sarawakians have voted in direct legislative elections on two levels: state and national, in single-mandate districts. The term of the parliament and state assemblies is five years. While in other Malaysian states the state and general elections take place simultaneously (it is a matter of practice, although not of legal obligation), in Sarawak this is not the case. Since 1978, when a general election took place and the state election in Sarawak was postponed until 1979, the state assembly follows its own five-year cycle.[30] It is the prime minister's prerogative to call for general election, although formally it is the Yang di-Pertuan Agong who dissolves the assembly prior to the next election, and the chief of the Election Commission who declares the nomination and polling dates. A corresponding setting exists for state elections, with the chief minister (or Menteri Besar), Yang di-Pertua Negeri and the state chief of the Election Commission carrying out the respective roles.

In 1964, elections to local councils were suspended because of the Malaysian–Indonesian "Confrontation"[31] and the resultant threat of instability. The suspension was never lifted, and in 1976 the Local Government Act replaced it (Rahman 1994, 10). Based on the Act, the councillors are nominated, while traditional local leaders (heads of longhouses, i.e. traditional dwellings, and villages (*tua kampong*), as well as Chinese Kapitans), previously elected by their respective communities according to their traditions, were from now on to be no more than civil servants, nominated and released at the government's prerogative. On the federal level, the Senate (or Dewan Negara) is a non-elected body consisting of members nominated partly by the Yang di-Pertuan Agong, and partly by the states (via state legislative assemblies), for three years. Although it is entirely a "rubber stamp" (Chin 2002, 211) body, nominations to the Senate are of political weight and carry importance in terms of ethnic proportions that need to be observed in a consociational state.

Before the second election in Sarawak could take place, new constituencies had to be drawn to comply with the nationwide electoral system. The Election Commission set off to delineate the new constituencies in 1965; administrative boundaries were not to play any role in designing the new constituencies. Generally, the Election Commission's delimitation exercise of 1965 did not substantially change ethnic strength in terms of seats in comparison to the 1963[32] situation. The Chinese remained starkly underrepresented, thanks to the constitutional allowance of a rural constituency containing "as little as half of the electors in any urban constituency" (Leigh 1974, 124). The Chinese vote, concentrated in urban centres, was bound to weigh less than the rural vote (chiefly Malay, Melanau and non-Muslim indigenous voters). However, says Leigh, in some constituencies in

Sarawak the disparities were much greater than constitutionally allowed (1974, 124). These limitations of disproportionality were entirely removed in 1973 (Chin 2002, 213).

The Election Commission carries out a delineation exercise every eight years: it makes readjustments to the shape of the constituencies and creates new seats. All changes need to be consulted with the local populaces, albeit not bindingly. Results of the exercise need to be accepted by the parliament and the state assembly, respectively; a two-thirds majority is required (hence the enormous pressure on the ruling coalition to maintain the said majority). The 1965 delineation exercise in Sarawak was crucial, as it was the first that established the constituencies not according to administrative units. Parliamentary seats were created by combining two state seats into one parliamentary one. "The constituency boundaries appear to have been drawn primarily with a view to the desired *state* representation, and little attention was paid to the political effects of joining each pair of state constituencies to form one parliamentary seat" (Leigh 1974, 134, emphasis mine).

Since 1990, each parliamentary (national) constituency for general elections includes two or three state seats in its boundaries. Prior to the biggest re-delineation exercise carried out by the Election Commission between 1984 and 1986, there were 24 parliamentary seats and 2 state seats for each of them (a total of 48 state seats) in Sarawak. The first general election after the re-delineation took place in 1990; the number of parliamentary seats for Sarawak came up to 27, for the next election (1995) to 28 and finally to the current 31 in 2008. For the 1991 state election the number of seats had increased from 48 to 56. The mid-1990s re-delineation increased the number of state seats in Sarawak to 62 (1996 election) and 71 (2006 election onwards). The 2015 re-delineation will bring the number of state seats to 82 (*Malaysiakini* 2015). The number of parliamentary seats coincides with the number of districts (administrative units) in Sarawak, but the boundaries do not. The Election Commission can shape the electoral districts to its liking. Creating new seats does not happen through a simple split of one existing seat into two; new seats are carved out freely from any existing ones.

Ethnic composition of seats changes significantly with every re-delineation and seats are renamed correspondingly. With the original distribution (1970–1987 elections), the non-Muslim indigenous peoples were overrepresented in the state assembly, while the Chinese were starkly underrepresented (compare Tables 2.2 and 2.3). After subsequent re-delineations, the Chinese remained underrepresented, however this time to the benefit of the Muslim voters. The non-Muslim indigenous currently have the number of seats that corresponds to their proportion in the society. Therefore, the 1969/1970 general and state elections have particular value for this study. They were conducted according to the FPTP single-member constituency principle for the first time, but were not burdened by the rigid rules of the permanent coalition agreement which precluded parties from competing against each other. Parties were also not limited in their choice as to which constituencies to contest: only the Bumiputera party and SCA agreed not to field candidates against each other, which they would be unlikely to do in any case, given their exclusive electoral base. Moreover, based on the result of the election,

Table 2.2 Ethnic composition of Sarawak 1970–2010

Ethnic groups	Muslim Bumiputera	Malay	Melanau	Non-Muslim Bumiputera	Iban	Bidayuh	Other/Orang Ulu	Chinese	Total Population
1970	24%	19%	5%	44%	31%	8%	5%	30%	975,918
1988	27%	21%	6%	43%	30%	8%	5%	29%	1,801,100
2010	28%	23%	5%	43%	29%	8%	6%	23%	2,471,140

Source: Year 1970: Leigh (1974); 1988: Jayum (1993, 79); 2010: State Planning Unit (2010, 10–13).

Note: Percentages do not total 100 because of non-Malaysian citizens.

Table 2.3 Seats in the Sarawak state assembly by ethnic majority

Ethnic groups/state election	Muslim indigenous	Non-Muslim indigenous	Iban	Bidayuh	Other/Orang Ulu	Chinese	Mixed	Total seats
1969–1987	13 (27)	25 (52)	17 (35)	5 (10)	3 (6)	8 (17)	2 (4)	48
1991	18 (32)	24 (43)	16 (28)	5 (9)	3 (5)	11 (20)	3* (5)	56
1996, 2001	23 (37)	26 (41)	18 (29)	5 (8)	3 (5)	12 (19)	1 **(2)	62
2006, 2011	25 (35)	30 (42)	20 (28)	6 (8)	4 (6)	13 (18)	3***(4)	71

Source: Years 1969–1987, 1991: Chin (1995); 2006–2011: The Star 2011; 1996: author's own compilation based on analysis and comparison between 1991 and 2006.

Note: In brackets, percentage of the total seats.

a new ruling coalition could be found between all possible combinations of parties, merely relying on their number of seats.[33] Therefore, the big unknown was also who might become the chief minister after the election.

Intense manoeuvring towards consolidation of the Malay/Melanau category by the Melanau leaders was under way. Prior to the 1969/1970 election, Minister of Education Rahman Yakub was known for his decisive push towards introducing Malay as the compulsory medium of instruction starting from first grade (Mauzy 1985, 161); this could hardly make him popular in his home state among the non-Muslims (Milne and Ratnam 1974, chap. 2). In his party's policy (whether BARJASA or Bumiputera), it was clear enough that the leadership was willing to go a long way to deemphasize the Malay-Melanau divide and focus on maintaining the image of the party and its leadership as part of the broader concept of Malays united not only between Malays and Melanaus in Sarawak, but also with the Peninsular Malays.

By 1968 the situation within the Alliance was far from clear as to the question of *who represents whom*. Given Pesaka's claim that as a native party it also represented the Malays, it clashed with the Bumiputera party. After the merger of BARJASA and PANAS, Bumiputera's main objective was to strengthen its position as the sole and united representative of the Muslim vote. In late 1968 Bumiputera pledged to Pesaka, the other native party within the coalition government, to surrender all its claims to field candidates in Muslim areas in the upcoming elections (Leigh 1974, 95). By this time, however, to fight for Muslim support was not enough: Bumiputera had already envisaged its struggle to unite the Malays and the Bidayuhs. The mathematics of such a party would be: 19% (all the Malays), 5% (all the Melanaus) and 8% (all the Bidayuhs). Coalition with the weaker Chinese (Foochow) SCA could bring the popularity of such a party up to more than 40%, if half the Chinese opted for Alliance-member representation, based on the merit of its strong ties to the federal government. This way, the most dangerous category, the Iban, would have been excluded and neutralized. The tactics of the Bumiputera party to control at least some quarters of the non-Muslim natives within the party was therefore not devoid of mathematical sense. However, within the Alliance, it was Pesaka that claimed to represent all the "Dayaks" and felt threatened by the Bumiputera party's strategy. "To counter this seeming threat, Pesaka opened recruiting in Malay-Melanau territories" (*Far Eastern Economic Review* 1970).

However, Pesaka was losing ground amongst its traditional supporters in the Rajang river area; some of Pesaka's electorate felt that Pesaka sold out to Bumiputera and did not truly represent the non-Muslim indigenous interests (Leigh 1974, 115–116). At the same time, SNAP, now a staunch opposition party, was increasing its standing not only among the non-Muslim indigenous, but also among disillusioned former PANAS members as well as Pesakas and was becoming a truly multiethnic party (Leigh 1974, 107). SNAP won four district council by-elections in 1967 (Leigh 1974, 115), and the party was clearly gaining support among Sarawakians who saw the 1966 federal intervention as undue. SUPP, on the other hand, as much in opposition as always, in 1969 finally declared that it "supported the concept of Malaysia" (Leigh 1974, 130), which had the effect that the party could freely campaign before the upcoming election without being accused of subversion.

Given strong opponents, SUPP and SNAP, it was crucial for the ruling coalition not to allow its components to compete against each other, but Pesaka decided to go against Alliance partners in the upcoming elections and contest 17 state and 7 federal seats against the Bumiputera party and SCA candidates (*Far Eastern Economic Review* 1970). The Alliance's chances in the upcoming election looked bleak. *Far Eastern Economic Review*'s prediction of the results was: "SNAP has the best chance of winning a simple majority in the Council Negri although SUPP will be a tough opponent in the Dayak areas where it has spent a lot of money and effort" (*Far Eastern Economic Review* 1970). Reece's optimism about SNAP's performance in the election and his expectation that this party would be the back-bone of the coalition (whether as SNAP-Pesaka-SCA, or SNAP-Bumiputera-SCA, or SNAP-SUPP) went as far as to toy with the idea of Stephen Kalong Ningkan returning as the chief minister (*Far Eastern Economic Review* 1970).

In May 1969, elections commenced across the Federation. However, because of riots in Kuala Lumpur on 13 May 1969, the polling in Sarawak was suspended, a state of emergency was introduced and the electoral process was not resumed until June 1970. The final results of the elections were as follows: SUPP gained 12 seats, Bumiputera 12, SNAP 12, Pesaka 8, SCA 3, Independent 1.[34] "Of the 28 Dayak seats 19 were won by mainly-Dayak parties, SNAP and Pesaka. Another seat was won by an Independent who later joined Pesaka. Out of the 12 Malay seats 10 were won by Bumiputera, and 1 more was won by the SCA, largely through Malay votes, provided courtesy of Bumiputera. All 8 Chinese seats were won by SUPP or SCA" (Milne and Ratnam 1974, 114, original spelling). SCA's three seats, of which two were won on the back of Malay support for the Bumi-putera party, indicate not only the party's lack of standing in general, but also that it could not be seen as a representative of the Chinese. The Chinese vote went almost entirely to SUPP. Therefore, SCA was by then a mere hostage to its Bumi-putera coalition partner.[35]

2.7 The second critical juncture: 1970 coalition negotiations

In 1970 coalition opportunities were numerous, and the question of which parties would be included in the cabinet was as important as the question of which would be excluded. The previous cabinet arrangement between three parties (Pesaka, Bumiputera and SCA) would be short of a mere two candidates and, it seemed, it would have been easy enough to persuade two representatives to cross over from other parties. However, Pesaka insisted on nominating the chief minister (by virtue of continuity), which was not acceptable for Bumiputera because the party won more seats than Pesaka. The quarrel over the chief minister's position opened the floor for the hitherto opposition parties, SNAP and SUPP, to join the negotiations.

SNAP, Pesaka and SUPP were just about to sign a coalition agreement when SUPP was found to be negotiating with the Bumiputera at the same time, after it received a signal from the Kuala Lumpur Alliance leadership that the emergency rule would not be lifted if Bumiputera were not a part of the ruling coalition in Sarawak. Therefore, SUPP knew that if it wanted to be a part of a winning coalition, it would have to be with Bumiputera, and so the party decided to enter a coalition with Bumiputera instead of SNAP and Pesaka. SUPP insisted on signing a binding document stipulating the conditions of the coalition. The conditions were that next to a Bumiputera-nominated chief minister there would be two deputy chief ministers, one nominated by SUPP and one selected from among the elected representatives of "the Iban race" (Leigh 1974, 144). Other points of understanding secured that SUPP would be consulted in all relevant matters. Although not included in the document, there were two more conditions. First, SUPP would not join the Alliance, and second, the party required that SCA would not be a part of the new deal. Rahman Yakub, a Melanau from Bumiputera, became the third chief minister of Sarawak.

SUPP's position on the federal level was at first undecided, as the party was not a member of the Alliance, but not long after Ong Kee Hui, SUPP's founder and leader, accepted a ministerial position in Kuala Lumpur. Consequently, the five MPs from SUPP were obliged to vote according to the government line. These five votes were critical for the Alliance to provide the two-thirds majority necessary to amend the constitution. Ironically, therefore, it was thanks to SUPP that the Constitutional Amendments Bill of 1971 could be passed; the bill "proscribed any further questioning of those provisions of the constitution relating to citizenship, Malay as the National Language, the special position and rights of the Malays and Natives, and the sovereignty of the Rulers" (Searle 1983, 141). In other words, the bill fortified everything that SUPP had been contesting during the previous years as an opposition party.

The Malay/Melanau-Chinese cabinet now needed non-Muslim indigenous members as tokens of consociationalism. Pesaka was officially still a member of the Alliance (definitely so on the federal level, where Pesaka's Temenggong Jugah held a ministerial portfolio), but the agreement between SUPP and Bumiputera meant that Pesaka was side-lined altogether and the party's grassroots were unlikely to accept a position in the cabinet that clearly would be without any bargaining powers. In the end, one Pesaka leader was coerced to join the cabinet and another young party member was persuaded to take up a ministerial position, while Temenggong Jugah upon his meeting with the prime minister was told that his portfolio in the federal cabinet would be given to someone else unless his party accepted the ruling coalition in its current shape and joined in. Jugah chose to obey.

Recalling these events, Jugah's son, Leonard Linggi Jugah (also a Pesaka leader), commented:

> In 1969. . ., 1971 . . . The emergency 1969, there was the election and there was chaos, no party has enough seats to form the government. In fact a lot of our people [were] saying that we should go out and be together and fight these people, but my father said, yeah, we should, but on the other hand if

we all go out we will force ourselves physically on Malays, our chances of succeeding [are] less. Why don't some of us stay? At least we can fight from within. That is the philosophy I followed when I stood up for election. Looking back, that was not wrong.

(Interview by the author, 21 October 2010)

In sum, contrary to all expectations from before the election, the Bumiputera party indeed prevailed as the game-setter in Sarawak after the election, although chiefly because of support from Kuala Lumpur. The coalition negotiations in 1970 constitute the second critical juncture in Sarawakian ethnic activation history. After 1966 it was known that the federal government would not hesitate to remove from office a chief minister who would not cooperate. From 1970 on it was established that a minimum-winning coalition in Sarawak is a coalition that has a majority in the state assembly and includes a party that represents the Muslim component – the latter condition guaranteed by the federal government again.

Finally, the consociational nature of the future cabinets was decided at this point. Bumiputera and SUPP discarded an option of coalition devoid of non-Muslim indigenous not because of the numbers in the assembly or the fear of riots (which, according to the Pesaka leader quoted earlier, were not merely theoretical), but, arguably, because it would undermine their chances in the next elections. Moreover, the tripartite division of ethnic categories materialized explicitly and parties – at least within the coalition – were assigned specific ethnic categories to represent: party Bumiputera the Malay/Melanau, SUPP the Chinese, and Pesaka the non-Muslim indigenous. Moreover, as Searle pointed out, "both those [Chinese and Malay/Melanau] communities had already realized their maximum political strength" (1983, 153). Bumiputera and SUPP as ethnic parties had already consolidated their control over the entire categories they were supposed to represent according to the West Malaysian standard of "one category, one party", or in other terms Bumiputera became the *titular party* for the *titular category* Malay/Melanau, and SUPP became the *titular party* for the *titular category* Chinese. The non-Muslim indigenous category and categories within it still offered numerous options for mobilization. However, the path on which ethnic relations in Sarawak would develop had been established by now and ethnic identity activation was bound to happen accordingly.

SUPP entered the coalition having set rigid conditions for Bumiputera. However, Bumiputera quickly managed to win over ten Pesaka and two SNAP assemblymen, and along with the three SCA legislators the coalition had 38 out of the 48 seats in the Council Negri. Therefore, even without SUPP, the Alliance would have a simple majority in Sarawak, which weakened the position of SUPP (Chin 1996a, 129). In 1974, among quite unclear circumstances, SCA dissolved and two out of its three elected assemblymen joined SUPP. The merger, however, did not indicate reconciliation of the Foochow and non-Foochow dialectical categories, nor was the merger a result of SUPP's courting of SCA. Dissolution of SCA seems to be the outcome of Chief Minister Rahman Yakub's manoeuvrings (Chin 1996a, 196). As a member of a multiethnic coalition SUPP understandably adopted a more conciliatory stance than it used to as an opposition party. As a consequence, it logically made itself vulnerable to outbidding strategies. In 1978, a group of Foochows (some of whom were earlier members from SCA) left the SUPP and in that year's parliamentary election[36] fielded three Chinese independent candidates against SUPP. The agenda was to challenge the party for not being Chinese enough. They "promised to seek recognition of degrees from Taiwanese and Chinese-medium universities, to safeguard Chinese education, provide more Chinese programmes on television and of course to fight for equal political and economic rights for the Chinese" (Chin 1996a, 156). SUPP replied by warning voters of "an anti-Chinese backlash and re-imposition of curfew"[37] (Chin 1996a, 157).

The challengers did not succeed but the outbidding strategy and SUPP's response reveal two important particularities of ethnic mobilization in the light of a multiethnic coalition. For one, a multiethnic coalition is in a position to present itself as a guarantor of peaceful inter-ethnic co-existence, and even if the chance of violence in a given case is very remote, the security factor may be deployed, and SUPP deployed this strategy by warning of a "backlash". Another particularity, shared by other parties in the Sarawakian ruling coalition, is that although they are representatives of their "assigned" ethnic categories (and not necessarily categories of choice, as in the case of SUPP), they cease to advocate for the interests of their respective categories, and instead promote the compromise that has been struck between the coalition partners.

The aforementioned Foochow-Sibu challengers sought to find an alternative way to undermine SUPP. Along with other SUPP discontents, they decided it would be easier to fight SUPP being part of a greater, Malaysia-wide organization (Chin 1996a, chap. 7). In 1978 they initiated the process of introducing the West Malaysian Chinese opposition organization called the Democratic Action Party (DAP) to Sarawak. DAP's agenda originated from the People's Action Party (PAP) of Singapore; after Singapore was expelled from Malaysia, DAP carried the banner of PAP in the Peninsula. DAP was and remains vocal in its criticism towards Malay privileges, Barisan Nasional's monopoly on power and electoral process abuse. With this programmatic stance, the party appeals mostly to urban Chinese and is seen as a Chinese party. DAP's leadership on the national and state levels is predominantly Chinese and the party contests the seats assigned to MCA in West Malaysia and SUPP in Sarawak. By this token it is a Chinese ethnic party. The first election for DAP to contest was the 1979 state election; DAP contested urban Chinese-majority constituencies and became a serious threat to SUPP. The subsequent fierce DAP–SUPP electoral competition made SUPP even more focused on its Chinese electorate; from then on, the party was on defence against an outbidding-oriented competitor.

It was shown that during the first 10 years of party politics in Sarawak, "Saribas Iban" and "Rajang Iban" were frequently activated categories. Each category supported its own party and its own leadership and subscribed to different visions of politics. The Saribas Second Division Ibans saw themselves cooperating with the Chinese closer than with the Muslims, wanted "Sarawak for Sarawakians" and had a younger, better-educated leadership of self-made men. The Third Division

Rajang Ibans were more remote and less educated, and their leaders were legiti-mized by their traditional positions as *Penghulus*. They preferred closer ties with Malays, with whom they had few conflicts, not living in the same areas. This divi-sion was very clear until 1970, says Searle (1983). He argues that the remaining pockets of support for Pesaka in 1970 came not from Rajang Ibans as such, but from poorer Ibans who saw their chance in being loyal to the government and in this way securing assistance in the form of Minor Rural Projects, direct financial aid to localities distributed by members of the state executive. Similarly, those who had already received material assistance from the government felt obliged to return the favour and voted for Pesaka. Therefore, by the time of the 1969/1970 elections, and even more after it, Pesaka was merely *the party in the government*, not the *Rajang Iban party* (Searle 1983, 111–113).

In 1973 the Bumiputera and Pesaka parties merged to create a new party, Pesaka Bumiputera Bersatu (United Bumiputera Pesaka party, or PBB). The negotiations started in 1971 and in January 1973 it was declared that the two par-ties become one, with Temenggong Jugah of Pesaka taking up the position of the first president of PBB (Searle 1983, 151). Pesaka leaders justified the move with the argument that for Ibans and other non-Muslim indigenous it was crucial to advance in economy and education first, before they could fully compete politi-cally with the Malay and Chinese. The grassroots did not share this view. Many Ibans believed that:

> Pesaka's merger with Parti Bumiputera thus represented the final betrayal of Iban interests to those of the Malays. In the parlance of the longhouse, Pesaka had been "eaten" by Parti Bumiputera – it had merely become a Parti Bumiputera lackey. Pesaka could therefore no longer claim to represent Iban political, economic and most importantly, communal aspirations.
>
> (Searle 1983, 156)

SNAP, on the other hand, kept extending its support base, and not only strength-ened its image as an *Iban* party, "irrespective of *river*, *district* or *Division*" (Searle 1983, 175), but also as a party of Ibans who strove to regain what they believed was due to them: the position of the chief minister. Between 1970 and 1974 the SNAP assemblymen focused on the issue of native privileges. According to Searle, the party put a lot of emphasis on the fact that non-Muslim indigenous were granted the same privileges by the constitution as the Muslims (1983, chap. 8). Cases of non-Muslim natives being refused their privileges were numerous. With the newly introduced New Economic Policy (NEP) and its strong economic means of positive discrimination, the question as to whether non-Muslim indig-enous would now be recognized as Bumiputera was of practical and financial consequence. This problem became a focus of SNAP's activities.

However, as Leigh pointed out, in the 1970 election SNAP gained the most eth-nically diverse support (1974, 140). Not only was SNAP one of only two parties that had won at least one seat in all five Divisions (the other party being Bumi-putera). In addition, SNAP, contrary to Bumiputera, did not rely on a particular ethnic category for its wins. SNAP contested 47 out of all 48 seats in Sarawak and after SUPP joined the ruling coalition, SNAP was the only opposition party. In 1970 SNAP was what SUPP used to be in 1963; the two parties in the respective periods, although associated with a certain ethnic category, were also seen as universalistic-oriented opposition, committed to a cause or idea that went beyond particularistic ethnic interests. Both parties, when in opposition, sought to become a counterweight to the ruling coalition, or, in other words, to become a non-ethnic alternative to the consociational politics. Furthermore, SNAP was still advocating for the recognition of Sarawak's distinct position within Malaysia, as the London Agreement had guaranteed, according to which Sarawak joined the Federation. In particular, SNAP emphasized that the position of Islam as religion and the Malay language in the state should be maintained according to the spirit of the London Agreement, and not to the Constitution of Malaya from 1957 (Searle 1983, chap. 8). These policy points were a continuation of the Ningkan cabinet's firm stand against the federal government and could help the party win over voters who were not happy about the strong ties between the ruling coalition and UMNO. The role of the central government in installing the Rahman Yakub cabinet was not to be forgotten soon.

SNAP maintained the opposition face throughout the 1974 general election with excellent results. The party won 18 seats, against its 10 seats won in the previous election. The win was mostly at the expense of the Pesaka seats; the Pesaka wing of PBB won only three seats, against its 10 from 1970 (Searle 1983, 175). What turned out to be of extreme importance for the future, SNAP came out of the election as the single strongest party, having obtained 43% of the total votes cast (Searle 1983, 175). SUPP retained its share of seats[38] from 1970, but with a lower percentage of the popular vote (21% against 29% in 1970). SUPP lost "a major share of its Chinese support to the opposition SNAP" (Chin 1996c, 140).

Cabinet negotiations were not needed after the 1974 elections; Rahman Yakub remained the chief minister, PBB and SUPP were by now the only two parties in the government. In 1974[39] the two coalition component parties split the non-Muslim indigenous seats between themselves. Even before the election results from the remote seats (non-Muslim indigenous) were known, Chief Minister Rahman Yakub (PBB Melanau) announced his cabinet (*Far Eastern Economic Review* 1974d). Clearly, despite the negligible support of the non-Muslim indigenous for the coalition, he did not plan to negotiate SNAP's participation in the government at that time.

This election was therefore crucial for establishing the ethnic division of parties. SNAP did not win in all non-Muslim indigenous seats; several went to PBB and SUPP, and even after joining the coalition SNAP could not claim them. A trade of seats would have been possible, but SNAP won no Muslim seats and only one Chinese. Therefore, since then it is not for the ethnic outlook of SUPP, PBB and SNAP that seats are allocated to the parties; the 1974 election, which was the last one before the grand coalition was established, decided the seat allocation within the coalition. The non-Muslim indigenous seats continue to be split between parties that are otherwise committed to representing other categories.

In fact, from 1970 until now there were no coalition negotiations or re-negotiations. Some parties exited the coalition, others were co-opted, a new chief minister (Taib Mahmud in 1981) assumed office, but none of these events was accompanied by renewal of the coalition deal. Most significantly, after the 1974 election the ruling coalition was forced to look for a long-term solution for its non-Muslim indigenous component. The PBB-Pesaka wing and SUPP non-Chinese assemblymen were too few to be even symbolic representatives of the non-Muslim natives, while SNAP was potentially going to not only consolidate all the non-Muslims and non-Chinese, but also Muslims and Chinese who were discontent with the government. The 1974 election proved that defeating SNAP in an election was not going to be easy.

SNAP's situation also required a difficult decision: the party had 18 assemblymen and nine MPs, but had no access to funds which could be distributed in their constituencies. MPs and state assemblymen in Malaysia are routinely assigned money as a form of assistance to their constituents. Especially in rural areas, the Minor Rural Projects, as well as bigger projects such as schools, roads and water treatment plants, are withheld from constituencies whose elected representatives are in opposition. Moreover, the political reality was such that no matter how high the party's electoral win would be in the future, SNAP would not be able to create the state government on its own terms, knowing Kuala Lumpur's preference for a Muslim chief minister. Therefore, the pressure on SNAP to act beyond idealistic agendas and induce material development to its supporters was enormous.

In 1976 the party decided to join the ruling coalition. The arguments justifying the move were quite similar to those Pesaka raised before it entered the deal with the Bumiputera party: the non-Muslim indigenous community should not be left out of the government, as people needed material development, which could be delivered only through the state machinery. The new generation of SNAP leadership that took over from the old guards like James Wong Kim Min and Stephen Kalong Ningkan (who both surprisingly lost their seats in the election, one to SUPP, the other to Pesaka) was much more pragmatic and less emotional. To young, well-educated SNAP leaders like Leo Moggie and Daniel Tajem, the decision to join the cabinet was a matter of reason and negotiations. They requested that upon SNAP's entry to the coalition, the party would be given crucial portfolios which would enable SNAP to deliver development to rural areas. Although PBB welcomed SNAP in the cabinet, it stalled the process of cabinet reshuffle and the ministerial nominations were delayed by several months. In the end, the portfolios allocated to SNAP were not of any significance (Searle 1983, 189). SNAP was the last party to join the coalition deal and its bargaining power was as little as that of Pesaka in 1970, both parties being numerically unnecessary to support the coalition.

The Malay/Melanau category in terms of its political role was consolidated with the creation of the Bumiputera party. The existence of a party that contests and wins all the Malay/Melanau seats in the state was what the Muslim elites in Sarawak and in West Malaysia had been striving for. Kuala Lumpur lent its helping hand to the party and presumably was expecting a strong and reliable state government in Sarawak from now on. On its end, the Bumiputera leadership in the state ensured strengthening of the Malay/Melanau influences in the state. To serve the purpose of promoting Malayness, several constitutional amendments were passed in 1976 (Searle 1983, 191). The timing was all but coincidental; only with SNAP in the coalition did the government have the two-thirds majority necessary to amend the state constitution. According to the Amendment Bill, the name of Council Negri was changed to Dewan Undangan Negeri. The new name in Malay stood for "State Legislative Assembly" and the form corresponded to the West Malaysian name of the state assemblies. The title of the governor, hitherto officially referred to as "Gabnor", was changed to Yang di-Pertua Negeri, which stands for the Malay "Head of State" and is a West Malaysian form. The most important amendment pertained to religious affairs of the state. Sarawak so far had had no head of Islam and Chief Minister Rahman Yakub pushed through a bill that made the Yang di-Pertuan Agong (i.e. the one of the nine Sultans from West Malaysia that is currently elected to serve as the "king") the head of Islam in Sarawak (Searle 1983, 191). This change corresponds to the legal solution adopted for the states of Penang and Melaka that do not have a ruler (also known as sultan), and the Yang di-Pertuan Agong serves as the Islamic head for them.

The Malay/Muslim image of Sarawak seemed now secured. However, to take the Malay/Melanau category for granted would be a mistake; there are, after all, "Malays" and "Melanaus" in it. Political reality informs that "Melanau/Malay" (as Chin (2004) refers to it) is a more accurate term for the category and discontent among the Malays would be understandable: the first two chief ministers of Sarawak were Ibans; the third and fourth were Melanaus. The Parti Anak Jati Sarawak (Children of Sarawak Identity Party or PAJAR) and its leader, Alli Kawi, a Kuching Malay, were the first to contest the Melanau leadership of PBB and, more importantly, oppose the chief minister's position being in the hands of a Melanau (Chin 1996a, 152). The party, established in 1978, was extremely short-lived and did not manage to win a single seat, but it did articulate claims of an otherwise dormant "Malay" category. The party "in fact clearly reflected this consciousness of the First Division Malays regarding their ethnic purity" (Saib 1985, 128). The party failed to mobilize any significant electorate, and Alli Kawi soon returned to the PBB camp (for Alli Kawi's account of the events, see Alli Kawi 2010), but the line of thought did not die.

In 1981 Rahman Yakub, a Melanau Muslim from the Bumiputera party, was replaced by his nephew Abdul Taib Mahmud, another Melanau Muslim from the same party, in the position of Sarawak chief minister. Taib Mahmud had been the puppet master of the Tawi Sli government in 1966 and, during that time, he left an impression of being a skilful leader. Moreover, Taib was well connected in Kuala Lumpur, having served in the federal cabinet. During his tenure as chief minister, Taib's patrimonial style of governing and his visible attempts to single-handedly control the state raised the objections of his uncle, Rahman Yakub, who became the Yang di-Pertua Negeri in 1981. A few years into Taib's tenure Rahman Yakub established a new party, Persatuan Rakyat Malaysia (Union of the Malaysian People, or Permas), to challenge PBB and Taib Mahmud. What on one hand looked like a personal or family dispute can also be expressed in ethnic terms.[40] Chin maintains that "in 1987 *the Malays* decided to back Rahman Yakub and Persatuan Rakyat Malaysia Sarawak (Permas) when Rahman decided to challenge his nephew for his old job [the chief minister's position]" (2003, 214, emphasis mine). Permas performed poorly in the 1987 election, and there is little explicit or implicit information that would allow categorizing it as a "Malay party".

Chin's (2003) interpretation of the Permas episode seems to be based on an assumption that the Malays' dissatisfaction would lead the category to lend its support to any Muslim-led organization that would challenge the Melanau-dominated government. According to Chin's findings, yet again in 1988 "there has been a concerted effort by some [Sarawakian] Malay political leaders (including some inside the PBB) to bring UMNO to the state. [. . .] Many Sarawak Malays (and *bumiputera*) have, in fact, joined UMNO although they are categorized as members of UMNO branches outside Sarawak" (2003, 214, spelling in original). Chin quotes a whopping 38,000 UMNO members in Sarawak in 1988, whose membership in the organization was supposedly dictated by disappointment with the Melanau-dominated leadership in the state (2003, 225). The Memorandum of Understanding between Taib Mahmud and UMNO, which stipulated that the organization would not establish branches in Sarawak (Chin 2003, 214), is a significant fact in general, but the evidence is insufficient to see the Taib–UMNO understanding as an ethnic matter. In fact, UMNO's presence in Sarawak would destroy the entire patronage structure in the state, which would be equally disadvantageous for Malays and Melanaus. In any case, except for the PAJAR episode, political parties in Sarawak have not activated the "Malay" category.

Consequently, the Melanau category has been deactivated; in fact, it is difficult to find information about PBB candidates' Malay or Melanau background. Obviously, this is no secret to the voters, but it is indeed an extremely rare occasion in the government media to read of "Malay candidate", or "Melanau minister". Based on my study of newspaper coverage of the 2006 and 2011 state elections, as well as the 2008 general election, media reports are quite exact in providing most detailed information pertaining to the ethnic backgrounds of non-Muslim indigenous candidates, but the "Malay" and "Melanau" adjectives hardly appear in the pro-BN press. The independent media are for the most part based in West Malaysia and often lack the insight to pay attention to the finer points of ethnic cleavages in the state. Whether in the BN-controlled or independent media, seats and their constituents are always described as "Malay/Melanau", although the regional split between Malays and Melanaus informs which seats are Malay and which are Melanau. Nevertheless, candidates' ethnic categories are merely implicit, albeit obviously well known to all the voters.

Note that PBB finds it suitable to field Melanau candidates in Malay constituencies. Taib Mahmud's and his son's victories in the Malay parliamentary seat of Kota Samarahan are a case in point. All the more interesting, Malays are not fielded in Melanau constituencies, and when a Malay candidate, Hasbi bin Habibollah, was fielded in the Muslim-Kedayan constituency of Limbang in the 2008 general election, he won with only a very narrow majority of 52% against the opposition's Chinese candidate Lau Liak Koi.[41] The opposition coalition Pakatan Rakyat (PR) prior to the 2011 state election also concluded that a Malay candidate in a Melanau-majority constituency could not win. Having to field a candidate against the chief minister in his Melanau Balingian constituency, the party decided that any Melanau candidate would be better than the best Malay (Tian Chua, interview by the author, 14 April 2011).

The Bidayuh or Land Dayaks were not as quick in establishing parties as the Iban, but the category offered equal potential for ethnic mobilization. The 1969/1970 state election was mainly a battle between SUPP and SNAP – both deeply in the opposition at the time – in the Bidayuh areas: three seats went to SUPP and two to SNAP. In two (Lundu and Bau) of the Bidayuh-majority seats, SUPP fielded Chinese candidates; these were also the two winning ones. In the other three (Bengoh, Tarat and Tebakang), SUPP fielded Bidayuh candidates, who all lost to SNAP (who fielded Bidayuhs in all five seats in question). Therefore, even with SUPP in the cabinet, there were no elected Bidayuh representatives in the state assembly on the government side, and it seemed that SNAP would soon be able to extend its appeal to this section of the natives, despite the party's prominent Iban outlook and rhetoric.

In 1971 a Bidayuh MP from SUPP was forced to vacate his seat[42] and a by-election was declared. Winning the by-election was crucial for the Bumiputera-SUPP coalition, given the fairly poor showing of the Alliance parties in the previous polling, and the shortage of the ruling coalition MPs in the federal parliament. Therefore Bumiputera intensified courting efforts among the Bidayuhs. It managed to win over a SNAP Bidayuh state representative, Nelson Kundai Ngareng, from Tarat, who, as a member of the Bumiputera party was immediately, still during the by-election campaign, nominated the state minister of youth (Leigh 1974, 154). Another Bidayuh, Dominic Dago ak Randan, "was appointed to head the Public Service commission and constantly referred to as 'the Bidayuh National Leader'" (Leigh 1974, 150).

The coalition candidate for the 1971 by-election came, however, from SUPP, as the party had won the seat in 1970, although Bumiputera attempted to plant an independent and "steal" the seat from SUPP.[43] SUPP fielded a Bidayuh, who competed against a Chinese candidate from SNAP and in the end, SUPP and the Bidayuh candidate prevailed. Therefore, the activation of the Bidayuh category happened for the first time after the 1969/1970 election and was of the Bumiputera party and chief minister's doing. In 1979 in the Bidayuh Bengoh state constituency, an alternative scenario played out. In its previously won seat, SUPP fielded its Chinese stalwart and one of the party's founders, Stephen Yong Kuet Tze, in place of the Bidayuh incumbent. PBB unofficially backed a local independent, a Bidayuh, who campaigned explicitly on the theme of ethnicity (Chin 1996a, 164). Replacing the Bidayuh incumbent with a Chinese, claimed independent Wilfred Nissom, was an act of disrespect from SUPP towards the Bidayuh community. The Chinese SUPP candidate won, however (Chin 1996a, 164). Moreover, Stephen Yong Kuet Tze as a federal minister from SUPP also managed to win the Bidayuh-majority Padawan (part of which is the Bengoh state seat) parliamentary seat in 1986, at the apex of the Dayak mobilization campaign, which was based on the principle that Dayaks should vote for Dayaks in Dayak parties. This is one of many examples of patronage trumping ethnic voting.

The first party to specifically and exclusively mobilize Bidayuhs was established in the early 1970s and was called BISAMAH. In the 1974 state election the BISAMAH party contested in four (out of five then existing) Bidayuh-majority state constituencies, albeit unsuccessfully (Fistié 1976, 355). The Bidayuh party did not take off, possibly because of intense courting of the category by the government around the same time as the party tried to establish itself. The absence of a party representing the category is, however, not equivocal to the category being non-activated. The mechanism of activating categories beyond party politics will be discussed in a later chapter of this work, and Bidayuh remains an activated category in Sarawak.

In 1996 the State Reform Party (STAR) was established by a prominent Bidayuh, Patau Rubis. Chin maintains that "it was widely perceived to be a Bidayuh political vehicle" (2003, 219). The party indeed fielded candidates chiefly in Bidayuh seats (*Malaysiakini* 2004), but the explicit message of the president in his foreword (Patau Rubis 2004) appealed to "Sarawakians" and called for "Politics of Reconciliation and Integration", never mentioning the Bidayuh. STAR's youth wing chose as its slogan "Towards Bangsa Malaysia". STAR was therefore an ethnic Bidayuh party by the token of its area of contestation, but neither its implicit nor its explicit message supports that. The party never won a seat in an election.

Although the Dayak category had been present in both the vernacular and ethno-political discourse in Sarawak prior to PBDS, its activation materialized only with the establishment of the Parti Bansa Dayak Sarawak (lit. "Party of the Dayak People", or PBDS). The PBDS targeted an electorate no other than that SNAP and Pesaka targeted, but under different premises. In 1981 the SNAP election for party president was won by James Wong Kim Min, a seasoned politician and a wealthy Chinese businessman. Wong's election was in line with the party's multi-cultural outlook that was very prominent during its opposition years. However, several "second generation Iban leaders" (Jayum 1993, 14) of SNAP, most prominently Leo Moggie, felt that a Chinese should not be the president of a party that strives to represent the non-Muslim indigenous; after all, this was the portion of the electorate that SNAP had been assigned as a coalition member.

As a result of this personality and ideology crisis, in 1983 Leo Moggie established PBDS, which has ever since been associated with the concept of "Dayakism". PBDS's manifesto included no mention of "Dayaks" but nevertheless paid a lot of attention to the cause of land policy and agriculture (Ritchie 1993, 131–140), clearly topics of vital interest to Dayaks. PBDS's message was so strongly Dayak-oriented that the organization opened a door to accusations of being a "racialist party" (Chin 1996b, 520).

The party sought to capitalize on the numerical advantage of the non-Muslim indigenous in the state, unite them under the "Dayak" umbrella and lift them up from their purported status as "second-class Bumiputera". The party, a breakaway organization from a Barisan Nasional component, applied immediately to be accepted as a member of the coalition. SNAP understandably vetoed the admission and Taib Mahmud as the chief minister devised a compromise solution. PBDS became an affiliate member of the BN Tiga Plus (BN 3+) coalition. Two PBDS federal cabinet ministers (Daniel Tajem and Leo Moggie), who had been nominated to the cabinet as SNAP members, retained their portfolios as PBDS members.

The BN 3+ formula was no solution for seat allocation. Prior to the 1983 election, SNAP and PBDS could not find a compromise on seat allocation and the Barisan Nasional leadership decided to allow the two parties to contest against each other, using their own party symbols (and not the usual BN symbol). An open contest between these two parties also invited independents unofficially backed by other BN parties; the vote was bound to be split and just about any candidate had a chance to win in non-Muslim indigenous constituencies. SNAP also fielded a candidate in a PBB (Bidayuh) seat and another in a SUPP (Chinese) seat (Chin 1996b, 188). The results of the 1983 election suggest that PBDS miscalculated its chances. The party won eight seats (of the total 48 constituencies), only one more than SNAP. PBB won in 20 constituencies and SUPP in 13.[44] However, PBDS's and SNAP's seats put together were three fewer than the 1979 SNAP result (18 seats). PBB and SUPP through their "independents" both managed to increase their representation in the state assembly by one and two seats, respectively, all three seats being non-Muslim indigenous.

The situation changed by the 1987 election. PBDS had more time to work on its strategy and to convey its message more strongly to the electorate. The Malay party Permas led by Rahman Yakub was working in parallel to undermine the chances of PBB and the two parties entered the election as a coalition that offered prospects of unseating the entire ruling coalition of the day.[45] The Maju coalition, as the Permas-PBDS coalition was called, aimed at winning Muslim and non-Muslim indigenous seats and establishing a new BN cabinet with Rahman Yakub as chief minister. PBDS's leader, Leo Moggie, did not contest the state election (already being an MP) and with this, eliminated the possibility of a Dayak becoming the chief minister, even in the case of Maju's win in the election.

As it turned out, the Maju coalition lost in the polls because of the poor performance of Permas; PBDS won 15 seats (of the total 48) and was one of the two biggest parties in the assembly as PBB won equal number of seats (Chin 1995, 19–24). PBDS's win was again mostly at the expense of SNAP, which this time won only three seats. SUPP won 11 seats, three of which were in non-Muslim indigenous constituencies. Of PBB's 15 seats, 12 were in Malay/Melanau constituencies, and three in non-Muslim Bumiputera. Permas won five seats, three Malay/Melanau, one mixed and one non-Muslim indigenous.

PBDS was at the time a member of the BN at the federal level, and technically was a BN-accepted party, regardless of its status at the state level. Therefore, had PBDS been accepted as a regular coalition partner on the state level, it would have an equal claim to the chief minister's post as PBB had, however, the party was not accepted into PBDS until years later. In the course of the assembly's term, eight PBDS representatives crossed over to PBB. Nevertheless, the 1987 election was deemed "The Dayak awakening" (*Far Eastern Economic Review* 1987), and the understanding that "majority ethnic group becomes the largest party in Sarawak" (*Far Eastern Economic Review* 1987) reflected the attitudes of many in the state. Indeed, for the "Dayak" category – activated for the first time – this was an "awakening".

In the 1987 state election and in the 1990 general election, the BN propaganda pictured PBDS as racist and exclusive. Chin estimated that this image cost the party about 90% of its potential Chinese electorate (1996a, 246). By 1991 the party decided to "soften its image by fielding seven Chinese candidates, some in Dayak-majority areas" (Chin 1996a, 246). PBDS had established a Chinese Affairs Consultative Committee, but the party's constitution still did not allow membership for non-Dayaks. The Chinese could contest on the party's ticket, but could not be members of the party. This caution was presumably related to the party's own origins; PBDS broke away from SNAP over the issue of Chinese leadership of the party, unacceptable to the non-Muslim indigenous in the eyes of PBDS.

In 1990 PBDS announced its ambition to nominate the chief minister after the next state election (*Malaysiakini* 2002). Yet again, however, Leo Moggie, the party's only potential chief minister candidate, did not contest in the state election and could not be considered for the position even if PBDS won a majority in the assembly. After the 1991 state elections, when PBDS won in only 7 out of 34 con-tested seats, the party requested admission to the state Barisan Nasional. The party was joining as a component that was not necessary to maintain majority and there-fore had no say about the conditions of entry, much like SNAP in 1976. In fact, this was the last inclusion of any party to the state BN[46] and its weakest partner that joined unconditionally. Leonard Linggi Jugah (member of the PBB Pesaka wing) commented on the entry in the interview: "Because some of us [Dayaks are] in there [government], [it is] much easier for PBDS to come in. I was one of those [. . .] in the committee, who was trying to find ways to, how they come in, what are the issues, what are the conditions. [. . .] I was quite surprised [. . .] when they came in, when they feel opposition is no longer the course of action, they come in and say 'we have no conditions' [to enter the government]" (interview by the author, 21 October 2010). PBDS was accepted as a BN member only in 1994 (*Malaysiakini* 2002).

The compelling question at this stage was how different PBDS was to SNAP in terms of categories they were appealing to. On one hand, within the ruling coalition, the two parties were entitled to compete in the same non-Muslim and non-Chinese seats. On the other hand, however, the explicit message the parties conveyed was different. SNAP had developed from a Saribas Iban to an Iban party, and between 1970 and 1974 it extended its appeal beyond ethnic particularism to attract all the anti-establishment voters. PBDS attempted to establish itself as the proper representative of the collective non-Muslim, non-Chinese category called "Dayak". "Dayak" would have materialized as an activated category if PBDS had contested and won all (or the overwhelming majority) of the non-Muslim indigenous seats. Only if these seats, of which many were so far represented by

PBB and SUPP, were claimed by PBDS would the party become the *titular party* of the *titular category*. PBDS, however, managed to win chiefly against SNAP. The 1987 election was PBDS's best performance and the party captured 15 out of 25 Dayak seats. Each of the three BN parties retained three Dayak seats, and Permas won one.

Three issues arguably weighed against PBDS in the eyes of rational voters: foremost, the fact that the federal government would not accept a non-Muslim chief minister; therefore PBDS could not be the ultimate winner regardless of its electoral performance. Secondly, the experience of Pesaka and SNAP joining the coalition beyond the coalition's minimum-winning mark taught that the last component to join is the weakest. If PBDS was to join BN after the election, it would be again the last coalition partner and, arguably, the one with the least influence. Thirdly, Permas, PBDS's coalition partner, was too weak to compete with PBB in the eyes of the federal government. These facts hindered the successful activation of the Dayak category; not only had many voters calculated that "Dayak and Malay/Melanau" under Dayak leadership would not be a winning coalition, many PBDS elected representatives concluded the same and joined PBB during the term. Significantly, SUPP never considered shifting sides, confirming that once the ethnic bargaining went down a particular path, it was unlikely to change.

In the 2000s both PBDS and SNAP underwent severe internal turmoil; in both cases the underlying issue was leadership struggle, with the problem of growing Chinese influences on the party leadership. Both parties were consequently deregistered by the Registrar of Societies and the prevailing leader registered his faction as a new party. Parti Rakyat Sarawak (PRS, Sarawak People's Party) replaced PBDS in BN, while the Sarawak Progressive Democratic Party (SPDP) was registered in place of SNAP and took its place in BN. Both new parties were entitled to inherit not only the membership of their respective predecessors, but also the allocated state assembly and parliamentary seats in BN. SNAP was later re-registered and in 2011 competed in the elections as an opposition party.

2.8 Conclusions

Between 1960 and 1990 category activation happened mostly via parties which were struggling to establish their position against other parties as legitimate representatives of categories. Joining the ruling coalition meant that each member party was assigned a category to represent. The party-led mobilization resulted in the final activation of three titular categories by the 1980s ("Malay/Melanau", "Chinese" and "Dayak"). Parties outside the ruling coalition behaved in one of two ways; some chose the outbidding strategy (PAJAR, DAP) and sought to compete for the support of a particular category; others mobilized among all ethnic categories and challenged the paradigm of ethnic elite bargaining (SNAP and SUPP before joining the cabinet). However, respective voters also retained in their repertoires activated categories "Malay", "Melanau", "Iban" and "Bidayuh", as well as Chinese dialectical categories (which also tend to be geographically concentrated), which happened through labelling of legislative seats. Therefore, voters kept shifting between at least two categories, depending on the political context.

Two critical junctures established the path for ethno-politics in the state. Federal intervention was in the background of both junctures with simple practical results: firstly, the federal-level coalition lends a helping hand to the party or leader that offers the greatest prospects of smooth cooperation with the federal-level coalition; secondly, there can be no other chief minister in Sarawak but a Muslim. Moreover, Sarawak's consociational modus was developed during this time. However, against expectations, constituencies are not allocated to parties based on the parties' ethnic appeal, as would have been the case if the power-sharing bargain were to be between ethnic categories and their respective parties. When SNAP joined the ruling coalition, it became clear that it was not the parties that share power on behalf of ethnic categories. The assembly constituencies and their rightful party contestants are not a matter of negotiation and are an effect of a historical process. Seats are allocated to the party that won the seats in the previous election, regardless of the party's ethnic outlook. Therefore, the exact terms of the ethnic bargaining in Sarawak escape the simplicity of three titular ethnic categories and their respective parties participating in power. The question of who actually shares power remains to be answered.

Notes

1 Nevertheless, these are only some of many other possible sets of categories: religious divisions among the Chinese (Buddhists, Christians, Confucianists) and Indians (Hindus, Christians, Muslims), as well as the very prominent dialectical divisions among the Chinese are alternative potential categories to be activated.
2 Most notably in Hazis (2012).
3 See more about NEP in Jomo (2004) and Khoo (2004).
4 Cases of persons of mixed parentage (one parent being Chinese) being refused the Bumiputera status in Sarawak are not seldom and invariably stir vigorous public debate (*Malaysiakini* 2009).
5 The exact wording of article 3 is: "Islam is the religion of the Federation; but other religions may be practiced in peace and harmony in any part of the Federation."
6 The Sultanate of Brunei was originally included in the scheme but very early opted for a separate state.
7 Compare Leigh (1974, chap. 2). Singapore held a referendum in 1962 in which citizens were asked to state their preferences towards *conditions* of the merger; being *against* the merger was not an option in the referendum.
8 Joseph Tawie, interview by the author, 5 August 2010.
9 For details see Ongkili (1972; 1985, chap. 6).
10 Compare the situation of Muslim immigrants from the Philippines in Sabah and mass conversions in this state in Lim (2008, 114–117).
11 The Senate's position within the political system is weak; however, senators are eligible for nomination to the federal cabinet. Such nominations are very common; 12 cabinet ministers (as of March 2012) were senators (Office of the Prime Minister of Malaysia 2012).
12 Of the 65 members of the Senate, 40 are nominated by the *Yang di-Pertuan Agong*, 25 by state assemblies. Sarawakian assembly nominates two senators, the monarch another three from the state.

13 Although the idea never died; the issue of Sarawak's and Sabah's rights within the Federation and the two states' claim to their special position remains alive. Sabah's Jeffrey Kitingan is the chief figure who carries this topic.

14 As Roff noted, by late 1960 ethnographers were already suggesting that Melanau and Malay would merge as categories; the distinctions were seen as not prominent enough, while the census takers were believed to see fit to merge the category in the questionnaire in the foreseeable future (1974, 29). The same was expected to happen to the Bisayas, also indigenous converts to Islam (compare Roff 1974: chap. 2 fn 49.). Contrary to these expectations, Melanau remains a category not only in the census but also in day-to-day politics.

15 Known as the most vocal champion of Iban and Sarawak rights, Ningkan was partly Chinese (*Far Eastern Economic Review* 1970). A similar caveat of mixed parentage will accompany bios of many ethnic *cum* political leaders in Sarawak and Indonesian Kalimantan.

16 Throughout the text I use the current spelling "negeri" (meaning "state in a federation"), except when in reference to the former official name of the state assembly I follow the traditional spelling "negri".

17 The Council Negri was introduced by the Brookes as an appointed body; the British established elected district councils (Leigh 1974, 49).

18 The Council Negri additionally included three nominated members and three *ex officio* members; this puts the total number of assemblymen at 42.

19 Independents, although polled at 30% of votes in the 1963 election, were not to select their representative to the national assembly. For all the details see Leigh (1974).

20 For analysis of the 1969 election in West Malaysia, see Vasil (1972). Background and consequences of the riots are to be found, among others, in Kua (2007) and Teik (1971).

21 The informal institution of PM's approvals for state executive heads was established during Alliance times: "[A]fter both the 1959 ad 1964 elections in Malaya the person appointed to head the executive in each Alliance-controlled state in Malaya, the Menteri Besar, or Chief Minister, had to be approved by the (Alliance) Federation Prime Minister" (Milne 1967, 80).

22 Leigh (1974) maintains that Abdul Taib Mahmud is from the First Division; as far as I was able to establish, Taib was born in Miri (Fourth Division), raised in Mukah (Third Division) and partly schooled in Kuching (First Division), but his family hails from the Third Division. Therefore, although I quote the cabinet composition after Leigh, for all intents and purposes Taib Mahmud is considered here a "Third Division Melanau".

23 James Wong Kim Min was included in the cabinet only after the British intervened on the grounds that racial balance must be maintained (Leigh 1974, 82); interestingly, he took the "Chinese" slot, but upon his nomination as deputy chief minister he joined SNAP (Leigh 1974, 95) and in later politics was a heavyweight in the "Dayak" camp.

24 For the legal background of the so-called Land Bill crisis and the strategic manoeuvres Taib undertook in this context, see Milne and Ratnam (1974, 215–224).

25 Technically speaking, "Singapore's separation from Malaysia was effected by a constitutional amendment, which was passed in each house [Dewan Rakyat and Dewan Negara] without any opposing vote" (Milne 1967, 76–77).

26 Wong and Chin (2011) reach similar conclusions.

27 In the nine Malay states that have a ruler, the Menteri Besar (equivalent of the chief minister) must be a Malay, unless the ruler waives this provision (compare the case of Perak in 2008, *Malaysiakini*, 2008). No such requirement exists for the four states (Penang, Malaka, Sabah and Sarawak) without a ruler.

28 Tawi Sli was a SNAP member until very shortly before the Ningkan crisis, when he crossed over to Pesaka (Leigh 1974, 95).

29 The "Ming Court affair" in 1987 (see more in Ritchie 1993) was an attempt to capitalize on the potential for cooperation between the Muslims and Dayaks from Parti Bansa Dayak Sarawak, but not as one category of Bumiputera.

30 After the first elections in Sarawak in 1963, state elections were due by October 1968, but were postponed to coincide with the general nationwide elections in 1969. Naturally, the opposition claimed that the delay was due to the Alliance's expectation of losing (Milne and Ratnam 1972, 56).

31 For background information on the Confrontation (Indonesian *Konfrontasi*), see Ongkili (1985, 6–7). Mezerik (1965) offers an excellent selection of official documents pertaining to the conflict.

32 Upon presenting the outcome of the exercise in 1968, the Bumiputera party suggested that the borders be redrawn in two constituencies to secure a majority of Malays at the expense of the non-Muslim indigenous. The EC accepted the suggestion, but it did not give disproportionate advantage to Muslim voters. As it was, the non-Muslim natives were overrepresented by some 12% (seats to population), and with the two disputed seats the disproportion would reach 14% (Leigh 1974, 125–126).

33 The 1987 election was similar in the fact that it offered the potential of reshuffling parties in the ruling coalition, including replacement of the PBB with an alternative. It did not, however, involve inter-ethnic competition, which was the case in 1970. For details on the 1987 state elections, see later in this chapter.

34 For the exact results and analysis, see Leigh (1974, chaps. 4 and 5).

35 Indeed, Chin counts 15 Bumiputera seats on account of the three additional SCA seats (1996a, 123).

36 Although the parliamentary election was called that year, Chief Minister Rahman Yakub postponed the state election until 1979.

37 Curfews were introduced several times in areas of Sarawak during intense communist activities (compare Chin 1996a, 73, fn. 57).

38 This time SUPP, as a member of the coalition, was assigned seats it was allowed to contest, chiefly the ones the party had won in 1970. Remarkably, SUPP lost the predominantly Chinese seat of Kuching Timor, but won the Miri-Subis seat from SNAP (Chin 1996a, 140–141).

39 This election in Malaysia, the first since the establishment of Barisan Nasional, was more of a referendum; to accentuate the particularity of the Sarawak election, note that Sarawak was the only state with no unopposed candidates. In Sabah, 15 out of 16 parliamentary seats were won unopposed; across the country, 47 parliamentary and 43 state seats were uncontested. Sarawak and Penang were the main battlefields of this election. No wonder, therefore, that SNAP with seven MPs was the strongest opposition party in the parliament (*Far Eastern Economic Review* 1974b). The statement "I don't care if you ask me to vote for a block of wood, as long as it has the Barisan Nasional label, I will vote for it" (*Far Eastern Economic Review* 1974c) was true that year for West Malaysian voters, but did not become true for Sarawakian voters for another two decades.

40 The affair can be also explained by a financial factor; the chief minister was in a position to disburse and withhold timber licenses and this issue was in the background of the Rahman Yakub–Taib Mahmud dispute. More on this in Aeria (2002), Neilson Ilan Mersat (2005), Ritchie (1993).

41 Faisal Hazis pointed out this case to me; he attributed the relatively poor performance of Hasbi to the fact that the Muslims in Limbang are Kedayan, not Malays. However, the Muslims are only 37% of the constituents in Limbang, the rest being an almost even mix of Ibans, Orang Ulu and Chinese.

42 Prior to the election the party had made all its candidates sign an undated letter of resignation, addressed to the speaker of the respective assembly. The letter was then kept by the party in case an elected representative was to leave the party. In such an instance the letter would be automatically sent to the speaker, resulting in the vacating of the seat. This was the case of Siyium ak. Mutit, who decided to quit SUPP, claiming that the party had neglected him and his constituency (Leigh 1974, 154).

43 In the end the chief minister thought better of it and the independent quit the race (Leigh 1974, 150–151).

44 These numbers include independents supported by the respective parties. Voters were aware of the independents' affiliation, i.e. voted knowingly. The independents later on joined the party which unofficially fielded them. PBDS had one such seat, SUPP two and PBB one (Chin 1996a, 195).

45 Behind the election there was an attempted coup d'état, called the Ming Court affair. These exciting events are not informative of ethnic identity change. For details, see Ritchie (1993).

46 SNAP and PBDS were later replaced by their successor parties (Sarawak Progressive Democratic Party and Parti Rakyat Sarawak, respectively) under different names and leadership.

References

Aeria, Andrew. 2002. "Politics, Business, the State and Development in Sarawak 1970–2000". Unpublished Thesis, London: London School of Economics, University of London.

Alli Kawi. 2010. *Ming Court Crisis: A Close and Intimate Knowledge of the Crisis behind the Scene*. Petaling Jaya, Malaysia: Golden Books Centre.

Chin, James. 1995. "Sarawak's 1987 and 1991 State Elections: An Analysis of the Ethnic Vote". *Borneo Research Bulletin* 26: 3–24.

———. 1996a. *Chinese Politics in Sarawak: A Study of the Sarawak United People's Party*. South-East Asian Social Science Monographs. Kuala Lumpur, Malaysia and New York: Oxford University Press.

———. 1996b. "PBDS and Ethnic Politics in Sarawak". *Journal of Contemporary Asia* 26 (4): 512–526.

———. 1996c. "The Sarawak Chinese Voters and Their Support for the Democratic Action Party (DAP)". *Southeast Asian Studies* 34 (2): 387–401.

———. 2002. "Malaysia: The Barisan National Supremacy". In *How Asia Votes*, edited by John Fuh-sheng Hsieh and David Newman, 210–233. New York: Chatham House Publishers.

———. 2003. "The Melanau-Malay Schism Erupts Again: Sarawak at the Polls". In *New Politics in Malaysia*, edited by Francis Loh and Johan Saravanamuttu, 213–227. Singapore: ISEAS.

———. 2004. "Sabah and Sarawak: The More Things Change the More They Remain the Same". *Southeast Asian Affairs* 2004: 156–168.

Far Eastern Economic Review. 1970. "Polling Heads". 12 March.

———. 1974a. "Sarawak: Yakub's Election Coup". 30 July.

———. 1974b. "Malaysia: Razak's Overkill". 23 August.

———. 1974c. "Electing to Be Moderate". 6 September.

———. 1974d. "Razak's Frail Eastern Front". 27 September.

———. 1987. "The Dayak Awakening: Majority Ethnic Group Becomes the Largest Party in Sarawak". 30 April.

Fistié, Pierre. 1976. "L'évolution de La Vie Politique Malaysienne". *Politique Étrangère* 41 (4): 337–369.

Hazis, Faisal S. 2012. *Domination and Contestation: Muslim Bumiputera Politics in Sarawak*. Singapore: ISEAS.

Jayum, A. Jawan. 1993. *The Iban Factor in Sarawak Politics*. Serdang, Malaysia: Penerbit Universiti Pertanian Malaysia.

Jomo, S.K. 2004. "The New Economic Policy and Interethnic Relations in Malaysia | Publications | UNRISD". UNRISD.

Khoo, Boo Teik. 2004. "Managing Ethnic Relations in Post-Crisis Malaysia and Indonesia: Lessons from the New Economic Policy?" UNRISD.

Kua, Kia Soong. 2007. "Racial Conflict in Malaysia: Against the Official History". *Race & Class* 49 (3): 33–53.

Leigh, Michael B. 1963. "Sarawak: Focus on Federation". *The Australian Quarterly* 35 (2): 39–50.

———. 1970. "Party Formation in Sarawak". *Indonesia*, no. 9: 189–224.

———. 1974. *The Rising Moon*. Sydney: Sydney University Press.

Lim, Regina. 2008. *Federal-State Relations in Sabah, Malaysia: The Berjaya Administration, 1976–85*. Singapore: ISEAS.

Malaysiakini. 2002. "Leo Moggie, political survivor". *Malaysiakini Online*. www.malaysiakini.com.

———. 2004. "Sarawak's Star to contest seats in Bidayuh heartland". *Malaysiakini Online*. www.malaysiakini.com.

———. 2008. "PAS' Nizar is new Perak MB". *Malaysiakini Online*. www.malaysiakini.com.

———. 2009. "Fury over 'disowned Bumiputera' in Sarawak". *Malaysiakini Online*. www.malaysiakini.com.

———. 2015. "Decision on review of S'wak delineation soon". *Malaysiakini Online*. www.malaysiakini.com.

Mauzy, Diane K. 1985. "Language and Language Policy in Malaysia". In *Language Policy and National Unity*, edited by William R. Beer and James E. Jacob, 151–175. Totowa, NJ: Rowman & Allanheld.

Mezerik, Avrahm G. 1965. *Malaysia-Indonesia Conflict*. New York: International Review Service.

Milne, R. S. 1967. *Government and Politics in Malaysia*. Boston, MA: Houghton Mifflin.

Milne, R. S, and K. J. Ratnam. 1972. "The Sarawak Elections of 1970: An Analysis of the Vote". *Journal of Southeast Asian Studies* 3 (01): 111–122.

———. 1974. *Malaysia—New States in a New Nation: Political Development of Sarawak and Sabah in Malaysia*. Studies in Commonwealth Politics and History, No. 2. London: Frank Cass.

Neilson Ilan Mersat. 2005. "Politics and Business in Sarawak (1963–2004)". Unpublished Thesis, Canberra: Australian National University.

Office of the Prime Minister of Malaysia. 2012. "Cabinet members". www.pmo.gov.my/?menu=cabinet&page=1797.

Ongkili, James P. 1972. *Modernization in East Malaysia – 1960–1970*. London: Oxford University Press.

———. 1985. *Nation-Building in Malaysia 1946–1974*. Singapore: Oxford University Press.

Patau Rubis. 2004. "President Foreword 2004". http://starbangsamalaysia.wordpress.com/2010/12/19/president-foreword-2004-dr-patau-rubis/.

Rahman, A. Rashid. 1994. *The Conduct of Election in Malaysia*. Kuala Lumpur: Berita Publishing.

Riker, William H. 1962. *The Theory of Political Coalitions*. New Haven, CT: Yale University Press.

Ritchie, James. 1993. *A Political Saga: Sarawak 1981–1993*. Singapore: Summer Times.

Roff, Margaret Clark. 1974. *The Politics of Belonging : Political Change in Sabah and Sarawak*. Kuala Lumpur: Oxford University Press.

Ross-Larson, Bruce Clifford. 1976. *The Politics of Federalism: Syed Kechik in East Malaysia*. Singapore: Ross-Larson.

Saib, Sanid. 1985. *Malay Politics in Sarawak, 1946–1966: The Search for Unity and Political Ascendancy*. Oxford: Oxford University Press.

Searle, Peter. 1983. *Politics in Sarawak, 1970–1976: The Iban Perspective*. Singapore ; New York: Oxford University Press.

Siddique, Sharon, and Leo Suryadinata. 1981. "Bumiputra and Pribumi: Economic Nationalism (Indiginism) in Malaysia and Indonesia". *Pacific Affairs* 54 (4): 662–687.

State Planning Unit. 2010. *Sarawak Facts and Figures 2010*. N.p.: Chief Minister's Department.

The Star. 2011. "10th Sarawak State Election/Nomination List 2011". 7 April.

Teik, Goh Cheng. 1971. *The May Thirteenth Incident and Democracy in Malaysia*. Kuala Lumpur: Oxford University Press.

Vasil, Raj Kumar. 1972. *The Malaysian General Election of 1969*. Singapore: Oxford University Press.

Wong, Chin Huat, and James Chin. 2011. "Malaysia: Centralized Federalism in an Electoral One-Party State". In *Varieties of Federal Governance: Major Contemporary Models*, edited by Rekha Saxena, 208–231. New Delhi: Cambridge University Press (Foundation Books).

3 Ethnic identity change in a consociational polity

3.1 The rules of the game

In the previous chapter, I showed how the ruling coalition and opposition in Sarawak were changing in search of equilibrium. However, by the mid-1990s Sarawakian politics had stabilized and consolidated. The relations between the state and the Federation had been established, the composition of the coalition was secured and the most dangerous pockets of opposition were contained. BN activities in the state entered the phase of maintenance and upkeep. According to assumptions about consociational polities, at this stage the set of activated categories should be fixed and arrested, with few or no opportunities for appealing to alternative categories.

The rules of the consociational game in the electoral realm in Sarawak are as follows:

1   Each constituency is represented by an assemblyman who shares an ethnic background with the majority of his/her constituents. The categorization of assemblymen, constituencies and majorities in each seat is "Malay/Melanau" (rarely "Malay" and "Melanau"), "Chinese", "Iban", "Bidayuh" and "Orang Ulu". These are the five categories entitled to participate in power in Sarawak if legislative representatives are used as a measure. In a typical situation a voter votes in both state and parliamentary elections for a candidate of the same ethnic category, i.e. a state Bidayuh seat is part of a parliamentary constituency in which Bidayuh is also the majority.[1] Press reports habitually underscore the ethnic composition of each constituency according to this breakdown (*The Borneo Post* 2006c; *The Star* 2011a). To emphasize the point, the "Dayak" category is not activated on the constituency level in Sarawak.
2   The strength and influence of each ethnic category are measured chiefly by two factors: by the number of constituencies in which the category is a majority, and by the number and importance of executive positions held by the members of each category.
3   Some categories are or historically were overrepresented; others are underrepresented in the state assembly as well as in the national parliament. The

over- and underrepresentation in the assemblies is achieved through gerrymandering. Constituency re-delineation exercises and creation of new seats are easy ways of giving one category more or less influence than its proportion in the population. Frequent gerrymandering ensures the desired ethnic composition of the assembly.

4   Seats are assigned to Barisan Nasional–component parties according to fixed principles: Muslim-majority constituencies are represented by PBB, Chinese-majority seats are represented by SUPP and non-Muslim indigenous seats are split between PBB, SUPP, PBDS (later PRS) and SNAP (later SPDP) based on rules of inheritance and other elusive criteria.

5   BN-component parties do not officially compete against each other.

6   Opposition candidates are usually of the same ethnic background as the Barisan Nasional candidates; inter-ethnic competition is avoided.

If these rules were rigidly observed, ethnic politics in Sarawak would hardly be a topic for study. Within the structure captured in the six rules, the only variance in ethnic identity activation would be observed 1) because of the party system not being fully compatible with the split between titular categories, and 2) because of two sets of titular categories effective in Sarawak: one that divides constituencies, and one that divides parties. In reality, however, other, non-titular categories are activated on different occasions and by different means.

Within this stable institutional framework, the ethnic element in state politics has remained as compelling as before in the past two decades in Sarawak. Contrary to expectations elaborated in Chapter 1, in this chapter it will be argued that not only does ethnic activation in Sarawak happen via multiple institutions (legislative elections, executive nominations, political parties) and on multiple levels (constituency, state, federation), but also that – despite Sarawak being a consociational, elite bargain–based polity – voters retain several activated categories in their repertoires.

To support this argument, this chapter will focus on the three main areas of ethnic activation identified in Sarawak: political parties, executive positions and the constituency level of the electoral process. By further study of parties it will be shown that these organizations, although undoubtedly ethnic, find space and need to appeal to categories other than the ones assigned to them by the principles of the coalition agreement, and the BN structure itself is a forum of ethnic appeals, contrary to the premise that the coalition eliminates inter-ethnic competition.

In Sarawak the "Muslim Bumiputera", the "Dayak" and the "Chinese" are the primary categories entitled to share power based on party representation, but other categories, mainly "Iban", "Bidayuh", "Orang Ulu", "Malay" and "Melanau" remain activated on the occasion of constituency representation. Although party seat allocation is almost fixed, candidates' nominations are often sensitive matters and several ethnic markers come into discussion, activating categories that are rarely otherwise present in Sarawakian politics. Furthermore, a study of executive nominations is due as the ministerial and senatorial nominations represent the elite bargain in a consociational polity. The crucial question for the analysis is – which category of a nominated minister's repertoire is brought to the fore in the nomination? Is a candidate nominated as "Malay", "Malay/Melanau", "Muslim" or "Bumiputera"? In other words, as a member of which category does the representative become the minister? This is asked to establish whether nominations are another arena where alternative categories are activated.

Therefore, the aim of this chapter is to point out how categories other than titular are activated in a consociational polity, and how it is possible that within different institutions distinct sets of categories are activated. All this will shed a different light on the question of which categories, and how many of them, participate in power sharing in Sarawak.

3.2 How do parties share power and to whom do they appeal?

The previous chapter showed the evolution of parties' ethnic appeal and the process of coalition negotiation. These two issues do not, however, exhaust the scope of party-related questions in this study. Two elements need to be investigated: how parties within the ruling coalition can increase or decrease their share of power, and how their ethnic appeal is maintained in day-to-day politics of a multiethnic coalition.

If parties share power via ethnic categories, and the parties are simplistically classified as PBB = Muslim indigenous, SUPP = Chinese, SNAP/SPDP + PBDS/PRS = non-Muslim indigenous, Table 3.1 would depict the dynamics of power sharing by party/ethnic category across the consecutive elections. The "Chinese" and "Malay/Melanau" (also known as "Muslim Bumiputera") categories are solid by the token of being represented in the BN by a single party. Non-Muslim indigenous candidates are found in all parties. The final contest between parties over the non-Muslim indigenous seats took place in the 1980s, but ended in the mid-1990s without a clear-cut result. Therefore, parties' ethnic images are neither rigid nor straightforward, and their ever-fluid ethnic appearances are partly responsible for Sarawakians' retaining multiple categories in their repertoires.

PBB is shown to have systematically increased its allocation, chiefly at the expense of PBDS (and indirectly SNAP). The flow of seats from SNAP and PBDS to PBB happened in the process of the non-Muslim party's elected representatives crossing over to PBB during the term and in the next election being entitled to stand in their constituency again under their new party. Most significantly, eight PBDS elected representatives switched to PBB after the 1987 election, as they had no intention to be opposition representatives, and BN was stalling the decision to accept PBDS into the coalition. These facts speak mostly for the extreme patrimonialism of Sarawakian politics. By 2006, PBB was only one seat short of having a simple majority in the assembly on its own.

These gains profited the Pesaka wing; the Bumiputera wing could only increase its seats by adding new Malay/Melanau seats in the process of gerrymandering, as it had already been allocated all the existing Muslim seats. As Table 2.3 shows, the number of Malay/Melanau seats doubled between the 1979 and the 2011

Table 3.1 State assembly seat allocation within Barisan Nasional 1974–2011

Party/election	1974 (48)	1979 (48)	1983 (48)	1987 (48)	1991 (56)	1996 (62)	2001 (62)	2006 (71)	2011 (71)
PBB	32	18	18	23	32	30	30	35	35
SUPP	16	12	12	14	17	17	17	19	19
SNAP*/ SPDP	x	18	18***	11	7	7	7	9	9
PBDS/ PRS**			18***	In opposition	In opposition	8	8	8	8

Source: Author's own compilation, various sources.

In brackets: total number of seats in an election.

* SPDP replaced SNAP in BN after SNAP's deregistration in 2002, taking over all its seats. SNAP was re-registered in 2006 and contested subsequent elections but not as a member of BN.

** PBDS was established only in 1981. PRS replaced PBDS in BN after PBDS's deregistration in 2003, taking over all its seats. In 1987 and 1991 PBDS was in BN at the federal level, but in opposition at the state level.

*** SNAP and PBDS were allowed to compete against each other in all 18 constituencies SNAP won in 1979.

elections, and all these newly created Muslim indigenous seats were invariably allocated to the PBB Bumiputera wing. Currently, 25 out of 35 PBB seats are contested by the Bumiputera, 10 by the Pesaka wing.

SUPP experienced no dramatic increase of seats and its allocation has grown proportionally to the expansion of the state assembly. SUPP has not been a destination party for disappointed members of other parties. A third of its currently allocated seats are non-Muslim indigenous seats, mostly a result of the 1970 election when SUPP was freely contesting seats not bound by coalition deals. That year SUPP won several non-Muslim indigenous seats and later on could claim the seats on the merit of incumbency. If PBB's and SUPP's ethnic inclination was to be judged by their allocated constituencies' majorities, the first would be a "Muslim + non-Muslim indigenous" party, and the latter a "Chinese + non-Muslim indigenous" party. Simpler, there would be a "non-Chinese" party (PBB) and a "non-Muslim" (SUPP) party. Needless to say, however, the PBB is invariably perceived as the Muslim or Malay/Melanau component, and the SUPP as the Chinese or non-indigenous component.

SPDP (formerly SNAP) and PRS (formerly PBDS) are allocated only non-Muslim indigenous seats. Although both these parties once changed their names and leadership, they inherited the seats of their predecessors. This was the case of SNAP, when the party within weeks became SPDP, and PBDS, when it became PRS. Therefore, SPDP and PRS both inherited their supporters (i.e. their constituencies) from otherwise defunct parties. As all BN members share ideology, parties use the Barisan Nasional logo, but also have their own logos, which in the case of SPDP and PBB display ethnic features (see Appendix 1.). There is little to go by to establish what the parties' actual support would be on the free electoral market. Although Malaysian media venture to analyze the electoral performance of each component party, it can be only done against the backdrop of the opposition party's performance in any given constituency. Any statement comparing the performance between the coalition components is methodologically unsound. The last election when parties within BN in Sarawak competed against each other was in 1983, and even this was only in non-Muslim indigenous seats.

After multiple state and general elections, several party regroupings, dissolutions, seat-swapping, party hopping and constituency re-delineation exercises, parties' claims that they exclusively command the support of the voters in any constituency are not logically valid. Malay and Melanau seats may be an exception, as the voters never had a chance to vote for any other BN party but PBB.[2] Otherwise, parties' claim to represent any particular ethnic categories is a tautology: parties contest the seats in which they claim to have support, and they have support in the constituencies they contest. Nothing more specific can be said about parties' ethnic support. Parties' membership numbers are hardly reliable: membership drives expand their numbers, without representing real support.

At this point it may be useful to imagine the point of view of an average voter in any given constituency in Sarawak. Which party would she prefer to contest her seat? Because all the component parties claim to be multi-racial (except PBB, which excludes Chinese as members), the voter may not be bothered much about the parties' ethnic outlook, which she may suspect not to be sincere in the first place. However, it is in the voter's best interest that the candidate is as close as possible to the top leadership and stands a good chance to be nominated to a possibly lucrative executive position, which will give him/her access to pork barrel. As we have seen, it is the PBB candidates who are the closest to the leadership and therefore have the best chance for nomination to any given post. Why then should the voter prefer PRS, SPDP or SUPP to contest her constituency? Moreover, it is the "BN leadership", not an individual party, that decides the nominees for each constituency, so which party contests the seat is hardly an important matter. In any case it is mostly a non-issue, as each representative is likely to be returned in his constituency as many as six times; but in the case of new seats, seats whose representatives fell out of grace, died or moved to another constituency, there may be a fair share of squabbling for nomination between parties or particular candidates/ethnic categories.

SUPP

A multiethnic party, of which Barisan Nasional is a textbook example, is principally a party from which no category is excluded. Barisan Nasional, while being a registered party, is also an entity composed of four parties (in Sarawak), of which some are ethnically exclusive in their explicit appeal. Reconciling the ethnically exclusive appeal with the ethnically inclusive one of the coalition requires sophisticated techniques. In particular SUPP has been continuously put in a position where the party had to balance its exclusive and inclusive image. The party is *the* Chinese component of the Sarawakian Barisan Nasional. Some facts to this effect were presented in the previous chapter, and the following discussion on ministerial nominations will enlighten more on this point. On the other hand, SUPP contests both Chinese and non-Muslim indigenous majority seats,[3] which was shown to be the result of the party's original appeal to a wide, non-ethnic electoral base from the 1960s and consequently its inherited set of allocated seats.

The particular Chinese appeal of the party can be shown using the example of the SUPP 2008 general election campaign. The party was sensing that the Democratic Action Party had an edge in the urban Chinese areas, the core of the *particular* SUPP base. Already in the 2006 state election SUPP had lost eight of its seats (19 allocated seats in total), and of the lost ones all but one were urban, Chinese-majority constituencies.[4] Therefore, a day prior to the polling in 2008, the SUPP published a promotional text in Sarawakian dailies under the title "Our Sincere Appeal", which unmistakably targeted Chinese readers and voters. "Issues on land lease renewal [the ever-unresolved problem of the Chinese community – KP], Chinese education and various local concerns have been attended to and rectified by us. SUPP, as a component of BN, will continue to strive for the betterment of the Chinese Community and to attend to your needs", read the "Appeal". "DAP is asking voters to betray the Chinese Community: **CHINESE FIGHTING AGAINST THE CHINESE**. Their motive is to weaken our Chinese representation in the Government. [. . .] Chinese power is in your hands" (*The Borneo Post* 2008b; original emphasis and spelling).

There could not be a clearer message that SUPP is the party that champions Chinese interests, especially if understood in the very narrow terms of Chinese ministers in the cabinet. SUPP is not expected to (and simply cannot) champion Chinese interests if these are understood as designing policies to the advantage of the Chinese[5]; the spirit of Barisan Nasional precludes that. Championing particular interests in the Malaysian context can mean only distribution of patronage, and this – according to the "Appeal" – SUPP was willing to offer to the Chinese. SUPP can hardly threaten the non-Muslim indigenous voters that they would miss their opportunity for government representation by not voting for SUPP, as gross of the executive nominations for the non-Chinese is assigned to other parties. Although some non-Chinese SUPP elected representatives (for instance, Hollis Tini) have held executive positions, the Chinese SUPP members take precedence in ministerial nominations and other spoils of power. Patronage is distributed to the non-Chinese SUPP constituencies according to the standard BN-loyalty electoral scheme. Therefore SUPP does not appeal to the non-Chinese voters as *their* party, but as *a* BN party.

SUPP also balances the cleavages between different Chinese dialects which are yet another basis for ethnic identification in Sarawak. The dialectical cleavage in SUPP partly overlaps with the regional divide (Chin 1996, 268) and stretches over four decades. Originally, the division between different Chinese dialects was emphasized by SCA and SUPP competition in the 1960s. After the dissolution of SCA, Foochows managed to find a way to the top echelons of SUPP leadership, especially in the person of Wong Soon Kai as secretary-general of the party (Chin 1996, 269). George Chan Hong Nam, a Hokkien from Miri who gained prominence in the 1980s, also challenged the predominance of the Kuching-based Hakka, Chao-anns and Hokkiens (Chin 1996, 268).

Dialectical differences remain an issue for electoral nominations within SUPP. Prior to the 2011 state election, the Sibu state seats under SUPP were split between the Chinese dialect communities, in a vernacular called "clans", as follows: Bawang Assan seat – Foochow, Dudong seat – Hokkien and Pelawan seat – Henghua (*New Straits Times* 2011). A DAP Hakka representative held the Bukit Assek parliamentary seat. During the nomination period, the Hokkien community in Sibu expressed its concern upon learning that the incumbent Hokkien would not contest the Dudong state seat within Sibu. SUPP went against the Hokkiens' wishes and Dudong was contested by a Foochow, who lost to the DAP candidate, himself a Hakka.[6] In the end SUPP managed to keep only one of the four seats in Sibu (Wong Soon Koh's).

The 2011 party presidential candidacy of Wong Soon Koh (a Foochow) sparked controversies within the party and subsequently "a rumour that efforts were underway to jettison Foochow from the party leadership" (*The Borneo Post* 2011h). In the end Wong withdrew[7] from the presidential contest, leading to a split and a crisis in the party (*The Star* 2011c). Tellingly, the crisis within the leadership led to the election of the first-ever non-Chinese deputy president of SUPP. The party's leadership had always been entirely Chinese, until in 2011 Richard Riot, a Bidayuh, was elected a vice-president of the party (*The Borneo Post* 2011i). It is worth pointing out that there are three Bidayuh elected representatives of SUPP (Richard Riot – the vice-president of the party and MP, Ranum Mina and Jerip Susil, both state assemblymen). There are also two Ibans of SUPP in the state assembly: long-term representative (formerly also MP) Francis Harden Hollis, and controversial newcomer Johnical Rayong.[8] As of mid-2011 Bidayuh was the second strongest category after Chinese among all the SUPP elected representatives (state and parliamentary combined), and of the six SUPP elected representatives at that time, four were non-Chinese (*The Borneo Post* 2011g). In the 2014 general election, only one SUPP candidate secured a seat in the national parliament, and it was Richard Riot. Currently, courting the non-Muslim indigenous component of SUPP seems the party's only chance of political survival. In 2011 the Chinese vote went almost entirely to the opposition DAP and it is the Bidayuh, and to an extent the Iban, who can hope for additional ministerial nominations through the SUPP. Note that in PRS, SPDP and PBB's Pesaka wing, it is the Iban who lead the parties.

Since 2011 SUPP faced a new problem, well known to some of its partners in BN: the risk of deregistration by the Registrar of Societies on the grounds of irregularities in party leadership elections, as two factions of SUPP separately elected a party president (*The Star* 2011c). In 2013, the conflict between factions exacerbated and a splinter called United People's Party (UPP) was established with several elected representatives of the SUPP (*The Borneo Post* 2013c). According to a well-known scenario, a group of SUPP elected representatives established a new party and claimed that the party is a BN member and that the new party is now the proper incumbent party in the constituencies whose representatives belong to UPP. Although the chief minister has not announced its final decision as to which party (SUPP or UPP) will contest the seats in question (*Malaysiakini* 2014), it is likely that the splinter and old parties will contest against each other to – as the logic goes – let the voters decide which party they support.

To summarize, within BN, SUPP is put in a position in which it has to project three different images.

1   A Chinese party, ever since its entry into the coalition with the Bumiputera party in 1970, when – in the "spirit of the Alliance" – it was assigned to represent one ethnic component (Chinese) in the government.
2   A multi-racial party; this was the party's initial appeal upon its inception, and as such it does not maintain ethnic wings in its structure and accepts members of any ethnic background.
3   A *non-Muslim* party in terms of its elected representatives and the profile of constituencies it contests; since the 2011 state election and 2014 general election, if elected representatives are considered, SUPP is a non-Muslim Bumiputera party.

Only the first, the Chinese appeal of the party, translates into championing the interests of a category. The promises of spoils in the form of ministerial nominations are addressed only to the Chinese, but this may be changing along with decreasing Chinese support for the party. SUPP is, however, a good example of the complicated ethnic outlook of each party that is a member of a multiethnic coalition. Maintenance of a pure, single-ethnicity appeal becomes virtually impossible.

SPDP and PRS

The ethnic appeal of the two non-Muslim indigenous parties, SPDP and PRS, is not less complex and arguably more elusive. As parties of all the non-Muslim indigenous categories, PRS and SPDP must find a balance between representing each of the titular categories (Iban, Bidayuh and Orang Ulu). In 2008 SPDP stirred some disappointment on the occasion of a Bidayuh not being appointed as a minister, when an Orang Ulu from this party received the nomination (more on this case later). PRS had to tackle a sensitive situation when the party decided to replace an Orang Ulu MP with an Iban candidate in the Hulu Rajang mixed seat (a slight Iban majority over the Orang Ulu) (*The Borneo Post* 2012).

On top of these already complex issues, according to James Masing, one of the main contemporary Dayak and Iban leaders, numbers speak for opening up to other ethnic categories, as the non-Muslim indigenous voters alone do not secure political power:

> The market is 44 percent Dayak and 56 percent non-Dayak. But that 44 percent alone is not yours only, PBB [. . .] is looking at it, SUPP [. . .] is looking at it, SPDP [. . .] is looking at it. So four component parties are looking at that 44 percent, but we do not have any share of the 56 if you are Dayak based. So your market is limited and shared by four, while for multi-racial parties the whole 100 percent is for them to tap.
>
> (*The Borneo Post* 2006e)

Historically, however, the said parties, despite their commitment to the indigenous component, struggled to stay independent from influential Chinese stalwarts. James Masing was elected president of the PBDS faction along with Sng Chee Hua, a Chinese, as his deputy, and this faction was later allowed to register as a new party, Parti Rakyat Sarawak (Sarawak People's Party, PRS). The other faction was strictly for pure Dayak leadership, as Daniel Tajem put it: "On that, I don't compromise. We can play a secondary or minor role but the party must be led and controlled by Dayaks" (*Malaysiakini* 2003). Ironically, during the crisis within PRS between 2006 and 2009, James Masing was fighting for leadership in the party against Larry Sng, the son of Sng Chee Hua. Although Masing's faction had earlier prevailed against Tajem's pure-Dayak faction, in 2006 Masing admitted that he was in the wrong criticizing "Dayakism" during the PBDS crisis, and confessed that he had "second thoughts on multi-racialism and the manner in which multi-racial parties come about when it involves the Dayak community" (*The Borneo Post* 2006e).

Masing's pro-Dayak feelings grew stronger as the squabble with Larry intensified. By 2007 Masing was quoted as saying, "the new generation of Dayaks should have pride in themselves and not let others [i.e. Chinese] lead them, as in the past" (*The Borneo Post* 2007b). At the same time, the party's new logo entirely discarded the "Dayak" image (see Appendix 1.) While the original PRS logo was a copy of the PBDS symbol with strong Dayak folk elements, the new logo is devoid of any ethnic elements; its green background and yellow rice stalk suggest that the party could represent any agricultural/rural community.

Among the non-Muslim indigenous, the Iban outnumbered the other components significantly and over time dominated politics of the Dayak category. PBDS, although it explicitly championed the interests of the Dayaks, was led by Ibans. Moreover, PRS and SPDP are currently dominated by Ibans. Except for SUPP, the highest-ranking non-Muslim indigenous in all parties, including the Pesaka wing of PBB, are Ibans. This issue is not part of the official discourse and the Iban facade of the parties is not an openly discussed, let alone challenged, matter. On the contrary, Iban leadership of the non-Muslim indigenous parties is almost common sense in Sarawak, as the facts described later confirm.

In 2002, just prior to SNAP's deregistration, the party was on a desperate search for a new leader. The 80-year-old Chinese president James Wong Kim Min (his presidency was the reason Leo Moggie split from SNAP in 1981 and started PBDS) needed to be replaced, but few candidates were on offer. The party's number two leader was also aging and other candidates had to be considered. Among them were "Baram MP Jacob Sagan and Dr Judson Tagal, the state assembly representative for Ba'kelalan, [both of whom] belong to the minority Orang Ulu community. They may have difficulty in garnering support in the Iban-dominated Snap" (*Malaysiakini* 2002a, original spelling). To emphasize the point, this Dayak party was therefore accepting a Chinese president (Wong Kim Min), but an Orang Ulu was not expected to gain support of the party members. Official elections for the position did not take place, as the party was deregistered.

SNAP's deregistration was a consequence of a wealthy young Chinese member and MP from Bintulu, Tiong King Sing, being expelled from the party by James Wong Kim Min. Significant pockets of the party, led by William Mawan Ikom (Iban), disagreed with Wong's decision and defended Tiong's position in the party. When SNAP was deregistered and its successor, SPDP, was registered, it was William Mawan Ikom who became the president, despite being "perceived as a poor administrator. Also, his messy personal life does not endear him to many of his colleagues. However, his status as a state minister as well as being an Iban could win him enough support for lack of another choice" (*Malaysiakini* 2002a).

PBB

Party Pesaka Bumiputera Bersatu (United Traditional Bumiputera Party, or PBB) is clear about its ethnic outlook: it is the Bumiputera category. The party accepts only Bumiputeras and only contests seats in which either Muslims or other indigenous are a majority. The Chinese are not allowed to join PBB. The appeal and the numbers of the party are traditionally tilted towards the Muslims. Despite this association with Islam, PBB was the main benefactor when PBDS elected representatives (all of them Christians) decided to leave the party after the 1987 election and joined PBB. Similarly, when SNAP was deregistered in 2002 and its representatives, 10 in total, including both the state- and federal-level representatives, needed to join a BN component to be entitled to defend their seats, speculation was that they might want to join PBB (*Malaysiakini* 2002b). However, it seemed that "Taib [Mahmud, then the chief minister and PBB president] will not accept the whole lot into his party's fold. [. . .] To open doors to all 10 would upset the racial equilibrium in PBB as well as cement suspicions that the Snap crisis was engineered to benefit one party" (*Malaysiakini* 2002b; original spelling). After SNAP's deregistration most of its members and elected representatives joined a new party, PRS, which replaced SNAP.

PBB maintains its dual "Bumiputera" and "Pesaka" image in a very prominent way even 35 years after the merger. Leonard Linggi Jugah from Pesaka explained:

> During my time it was quite clear cut: what is due to us [Pesaka], we just put it, this is our [constituency] allocation, so they [the Bumiputera wing] don't touch it, for some reason. So we maintain our side of the bargain. So whatever we have to do we deliver [i.e. win the constituencies allocated to us].
> (Interview by the author, 21 October 2010)

Elections to presidential posts and the PBB Supreme Council are carried out separately for the two wings. It goes without saying that the president's position is always returned unopposed (*The Borneo Post* 2007a). One of the two deputy president and one of the two senior vice-president posts go to each of the wings (*The Borneo Post* 2007a). The seven vice-president posts are split 4 to 3 between Bumiputera and Pesaka, respectively[9]; Supreme Council memberships are split 11 to 9, also to the advantage of Bumiputera (*The Borneo Post* 2007a).

Indisputably, parties are autonomous in electing their own leadership. According to the law, they have to hold their Triennial General Assembly (TGA) and hold elections for the Supreme Council and other positions. In practice, the current leadership presents the assembled members a line-up of leaders, who are then elected uncontested for the position. Another bulk of party posts is filled via nomination from the president of each party. Several parties have a very turbulent experience with party elections; in fact, all the so-called Dayak parties (SNAP twice, PBDS, PRS and SPDP once) have struggled to maintain their internal unity. Holding two TGAs by two conflicting factions of a party has happened several times; on numerous occasions parties in Sarawak have had two elected presidents and it was left to the decision of the Registrar of Societies as to which faction and president were the rightful ones.[10]

Despite Barisan Nasional having a rigid structure and parties hardly having any margin available for ethnic manoeuvring, BN components do engage in ethnic deliberations. Internal leadership elections and attention given to particular ethnic categories can influence a party's image. Although BN components are not in a position to champion policy-framed ethnic interests, their strategic patronage disbursement can impact their overall reception in the society. Parties, therefore, were shown to enjoy a margin of freedom for the activation of categories different from the three titular ones, e.g. by promoting ministerial nominations, manoeuvring composition of party leadership or making public statements underscoring parties' commitment to certain categories. Therefore, parties alone are more flexible about their ethnic outlook than Sarawak's institutional background would suggest them to be.

3.3 Executive and senatorial positions as a field of category activation

The ministerial positions, on both the federal and state levels, are primarily divided between parties (by, respectively, the prime minister and the chief minister), based on parties' performance in the elections, but the nominees' ethnic background is also observed. Since 2006, there were 48 ministerial portfolios in Sarawak, held by 29 assemblymen (it is common for one assemblyman to hold multiple portfolios). The distribution of these among the parties and ethnic categories is part of the sensitive power-sharing scheme. Considered here is the composition of the cabinet in 2011, as the cabinet was sworn in after the 2011 state election, as the line-up was heavily impacted by parties' and candidates' showing in the election. Although in February 2014 Taib Mahmud stepped down as the chief minister (also as PBB president) and Adenan Satem, a long-term cabinet member and Taib's former son-in-law, took up the chief ministership, the composition of the cabinet did not change (*The Star* 2014), whether analyzed from the perspective of ethnic category strength or party strength.

In 2011, the state cabinet portfolios were split as follows: PBB held 31 (of which 23 are in the hands of Muslims: Malays held 16, Melanaus 6 and Kedayan 1); SUPP held 7, while the other three parties held 5 each. Among the 29 assemblymen who held the portfolios, the split was as follows: 13 Muslims (10 Malays, 2 Melanaus, 1 Kedayan), 14 Dayaks (10 Ibans, 3 Bidayuh, 1 Orang Ulu), 2 Chinese (both Foochow) (compare Table 3.2). This distribution and the striking weakness of the Chinese in the cabinet is a consequence of massive Chinese support for the opposition DAP that swept urban seats which were traditionally SUPP's. For the first time in Malaysian Sarawak's history, there was no Chinese deputy chief minister (which, interestingly, is a breach of the original 1970 PBB-SUPP coalition agreement), but all (in this case two) elected Chinese representatives from Barisan Nasional hold a ministerial portfolio.

The Chinese problem had already appeared during the 2009 cabinet reshuffle, when the chief minister nominated new assistant ministers from each party except for SUPP (*The Sun* 2009). After the 2011 state election, the problem became even more intense. There were altogether only two Chinese representatives elected from BN and SUPP, and the sitting president of SUPP *cum* deputy chief minister was not returned in his constituency. After the election the defeated SUPP attempted to discourage the chief minister from nominating any Chinese ministers; indeed, it called for Bumiputera-only nominations from the party: "SUPP Central Working Committee's (CWC) [expressed an] objection to Wong [Soon

Table 3.2 Sarawak executive composition by ethnicity, September 2011

Ethnic background	Cabinet members*	Cabinet and non-cabinet ministers	Portfolios
Malay	3	10	16
Melanau	2	2**	6**
Kedayan	0	1	1
Total Muslim Bumiputera:	**5**	**13**	**23**
Iban	3	10	16
Bidayuh	1	3	4
Orang Ulu	0	1	1
Total Non-Muslim Bumiputera:	**4**	**14**	**21**
Chinese	1	2	4
Total	**9**	**29**	**48**

Source: Author's own compilation based on *The Borneo Post* (2011f).

  * According to the Constitution of Sarawak, only the chief minister and ministers are members of the cabinet. Assistant ministers are mentioned in the constitution, while second ministers are not, but are widely considered members of the executive.
** Includes Fatimah Abdullah, a Chinese Muslim married to a Melanau.

Koh]'s re-appointment as a state minister and instead requested that the post be given only to its non-Chinese elected representative"[11] (*The Borneo Post* 2011b). West Malaysian UMNO-related press expressed a similar opinion:

> The average Chinese voters have rejected BN and supported DAP. There-fore the BN state government can no longer be too generous to give place to representatives from the community. Sarawak cabinet must be reflective of the decisions and attitude of the voters. [. . .] Clear message must be sent. [Chief Minister] Taib must show gratitude to those that supported him and BN government.
>
> (*Utusan Malaysia* 2011)

SUPP Chinese voters were therefore attacked from two corners: by UMNO in the Peninsula and by their own party within the government. Ironically, an Iban leader from PRS came to their rescue: "The Chinese community should be represented in the State Cabinet even though the Chinese-based Sarawak United People's Party (SUPP) won only two urban seats in the recent election. Parti Rakyat Sarawak (PRS) president Dato Sri Dr James Masing said the inclusion of Chinese elected representatives in the cabinet would reflect that Sarawak was a truly multi-racial state" (*The Borneo Post* 2011b). The chief minister's reaction was to nominate both elected Chinese representatives to his cabinet. "I promised the Chinese that I want to help them get fair representation so I take what I can get and after all Soon Koh (Wong) was voted on [the] BN ticket and not SUPP" (*The Borneo Post* 2011b).

The West Malaysian equivalent of SUPP, the Malaysian Chinese Association (MCA), saw the situation very differently and sought to remove Taib from his chief minister post for not securing support from all ethnic groups. MCA's "senator Gan Ping Sieu call[ed] for Taib to step down immediately for allegedly failing to secure the support of every race in the 10th state elections" (*The Borneo Post* 2011b). Another non-Muslim indigenous leader, SPDP's Sylvester Entri, came to Taib's defense: "We had won the war and only lost pockets of battle, thus it does not justify relieving the general of command. [. . .] The bottom line however is, he is not the chief minister for the Chinese only but also for other races in Sarawak" ((*The Borneo Post* 2011b).

In this context a PBB Supreme Council member and senator, Idris Buang, went as far as to suggest that the Constitution of Sarawak should be amended to allow nomination to the cabinet of persons who are not elected representatives (*The Borneo Post* 2011e), so that "the Chinese community which only has two representatives in SUPP now would be better represented as many of them [i.e. Chinese voters] still support BN". The chief minister discarded such a move (*The Borneo Post* 2011e), and instead nominated one of the Chinese elected representatives as a minister (with two portfolios), and another one as an assistant minister (also with two portfolios). Note that BN expresses deep concern about Chinese representation in the cabinet and among the BN member parties, but at the same time consistently precludes the creation of additional seats in urban areas, which would balance out the Chinese underrepresentation in the state legislative assembly.

Similar debates spark in the context of federal cabinet nominations. After the 2008 general election in which Muslim and non-Muslim indigenous Sarawakians proved the most consistent supporters of BN nationwide, a Democratic Action Party leader, Wong Ho Leng from Sarawak, called for an Iban deputy prime minister. An Iban DPM, goes the logic, would be a natural consequence of the Iban loyalty to BN. James Masing, one of the two highest-ranking Iban politicians in Sarawak, replied that "the matter (Wong's suggestion) is not practical taking into account the Ibans only make up less than three percent of the total Malaysian population. Secondly, the issue might be considered sensitive in Malaysia" (*The Borneo Post* 2008c). Wong's calculated move again put the Iban leadership in the uncomfortable position in which ethnic parties frequently find themselves when in a multiethnic coalition. Masing, in order to defend the power-sharing scheme, was forced to act against his community interests. That it is the Chinese opposition leader who advocates more political recognition for the Iban must be particularly disconcerting to the Iban leadership, and it is as paradoxical as an Orang Ulu politician advocating for more Chinese ministers.

Federal ministerial nominations are also accompanied by leaders of particular categories vying for more nominations for their categories. When the cabinet line-up was presented in 2008 by Prime Minister Abdullah Ahmad Badawi and there were no Bidayuhs among the nominated, the Bidayuh MP, Tiki Lafe, expressed his disappointment (*The Borneo Post* 2008d). At the same time, an Orang Ulu, Jacob Sagan, was nominated to the federal cabinet, and on the occasion of his appointment Sagan thanked his community for being patient – he was the first Orang Ulu to join the cabinet in 20 years (*The Borneo Post* 2008e). Tiki blamed his party, SPDP, for not doing enough for the Bidayuh (*The Borneo Post*

2008d). The party's president (an Iban, William Mawan Ikom) responded that "SPDP is a multi-racial party, but a small one in the Barisan Nasional, and thus it cannot just champion the cause of one particular race." On this occasion the media reminded readers that the new nominee, Jacob Sagan, is a Kenyah – an ethnic category within the Orang Ulu (*The Borneo Post* 2008d). After the 2013 general election, the federal cabinet ministers (both full and deputy) from Sarawak included one independent (Idris Jala, Orang Ulu); seven PBB MPs (four Malays, two Ibans and a Bidayuh); two PRS MPs (both Iban); an SUPP MP (a Bidayuh) and no SPDP MPs (although the party had four MPs in total) (*The Borneo Post* 2013a).

How is the power-sharing scheme reflected in the composition of the Sarawak cabinet if we assume that it is the parties (not ethnic categories) that share power? Table 3.3 shows the Sarawak cabinet composition by party after the 2011 state election. PBB dominates visibly, whether by the number of members nominated to the cabinet, or by the number of portfolios they hold. This is no surprise given the *primus-inter-pares* position that PBB enjoys; PBB contests (and invariably wins) in 49% of all seats. Its position in the cabinet is even stronger; almost two-thirds of all cabinet members are from the chief minister's party; PBB members hold both the chief minister and the only deputy chief minister positions. Immediately after the shocking news of George Chan's (the hitherto deputy chief minister from SUPP) defeat in the 2011 state election, the biggest question was who would replace him: a Chinese from SUPP or a non-Muslim indigenous from PRS or SPDP (next to PBB Iban Deputy Chief Minister Alfred Jabu)? None of these options suited the then chief minister, who decided to cut the second deputy chief minister position; Adenan Satem, the new chief minister, also left the position unfilled.

The three other BN components are assigned portfolios and nominations based on their performance in the election, or so it would seem. The dispute presented earlier pertaining to SUPP's fate after the election proves this rule to be far from rigid; although there is no more deputy chief minister from the party, the number of SUPP's members nominated to the cabinet evens PRS's share, despite PRS's better performance in the 2011 state election. At the same time, SUPP also holds more portfolios. SUPP's troubling participation in the cabinet after its electoral debacle emphasizes the difficulty of managing a power-sharing scheme in the light of the low popularity of one of the component parties. In 1970, when SUPP

Table 3.3 Sarawak cabinet composition by party, September 2011

Party	Seats in the assembly	By cabinet members	By portfolios held
PBB	35 (35)	18	31
SUPP	6 (19)	4	7
PRS	8 (9)	4	5
SPDP	6 (8)	3	5
Total	**55 (71)**	**29**	**48**

Source: *The Borneo Post* (2011f).

Note: In brackets number of seats contested by each party.

joined the coalition, it brought SCA to dissolution; SCA simply did not enjoy popular support. By this logic, the chief minister should now start coalition talks with DAP to make sure that the party that represents the Chinese in the cabinet actually enjoys the support of the Chinese electorate. Of course DAP's participation in any BN government is unthinkable, and so the consociational all-inclusive nature of the BN state government becomes wobbly.

Senatorial nominations are clearly conducted according to power-sharing rules, although here it is the parties that seem to be the agent that carries the share. The Web site of the Dewan Negeri (Senate) presents the background and party membership of the senators. In March 2012 the Sarawak state assembly nominated two PBB senators (both are Orang Ulu: Lihan Jok, from PBB, and Idris Jala, independent). Among the senators nominated by the king, there was one PBB member (a Malay, Dayang Madinah binti Tun Abang Haji Openg), one PRS (an Iban, Doris Sophia Brodie) and one SPDP (a Chinese, Pan Chiong Ung). After the 2013 general election, the Sarawakian senators were as follows: Lihan Jok (Orang Ulu, PBB, nominated by the state assembly); Idris Jala (Orang Ulu, nominated by the king, independent); Doris Sophia Brodie (deputy speaker of the Senate, PRS, nominated by the king); Dayang Madinah binti Tun Abang Haji Openg, (Malay, PBB, nominated by the state assembly) and Sim Kui Hian (Chinese, SUPP, nominated by the king).

Lihan Jok's nomination as a senator by the state assembly sparked discontent among the Bidayuh. After Jala, Jok was the second Orang Ulu senator in office (and the sixth in history), but there had never been a Bidayuh in the Senate (*Free Malaysia Today* 2011). A Bidayuh member of the Democratic Action Party, Edward Luak, promptly asked: "Why has the Bidayuh community been forgotten in this respect? It appears that the election or the appointment of senators in PBB is being alternated among the Malay, Iban and Orang Ulu communities. And why are the Bidayuh leaders in the BN, especially in PBB, not making any noise? Why are they silent on this issue?" (*Free Malaysia Today* 2011). Michael Manyin Jawong, a Bidayuh state minister and chief advisor of the Dayak Bidayuh National Association, replied, "We already have two deputy ministers (at the federal [level]), one full minister and two assistant ministers at the state level and one Deputy Speaker (at the State Legislative Assembly). [. . .] So, what else do we want as a small community? I think this is the best the Bidayuhs have had. At the moment, I don't think it is fair for us to keep on asking this and that. We have more than enough, actually" (*The Borneo Post* 2011e). There is nothing new either in an opposition Bidayuh politician's claim for perks for his constituents, nor in a government Bidayuh politician's answer. After all, a multiethnic coalition precisely serves the purpose of limiting particular communities' claims. BN politicians (with the exception of the UMNO) go actually as far as downplaying their own category's value and acting contrary to the community's interests.

The fact that even the minimally influential senatorship is a position that carries ethnic weight is an important finding. Senatorial nominations are, as it turns out, an occasion to mobilize categories, much like state and federal ministerial posts and constituency delineation results that decide how many representatives each titular category has. While based on the senatorial nomination, the impression might be that the power is shared among the parties, it is also in the end a matter of inter-ethnic balance.

3.4 Constituency-level category activation

As was pointed out earlier, the creation of new seats in ethnically mixed areas may lead to the emergence of complicated ethnic demographics at the constituency level. Candidate nominations and party claims to a seat can become hot issues, but most importantly, new ethnic categories may be activated. Three new seats in the 2008 general election caused some disagreements between PBB and SUPP. The ethnic background of the constituents was studied to establish which party had the right to contest the seat. In its effort to contest the Sibuti constituency, SUPP claimed that the proportion of the voters there was "40 percent Ibans and 30 percent each Malays and Chinese"; this proportion would seemingly support SUPP's claim. PBB responded that SUPP had outdated numbers, and according to PBB's numbers, the seat "belonged to PBB" (*The Borneo Post* 2008a). Consequently, the latter party "urged SUPP to stop lobbying for the Sibuti seat since BN component parties do not believe in stirring up sentiments among the grassroots, but in power-sharing and cooperation" (*The Borneo Post* 2008a). In the end, Sibuti was contested by PBB; the candidate was a Kedayan, a small category of inland Muslims.

Two mixed state seats, Bekenu and Lambir, illustrate the same issue. The Muslim indigenous are a plurality in these mixed seats, and among the Muslims, the Kedayan local category is dominant. Therefore, Kedayan becomes the minimum-winning coalition in these two seats. It is otherwise a category of less than 1% of the Sarawakian population, and as Kedayan would not be in a position to successfully contest in elections, not only because of its numbers, but because Kedayan fits within the Muslim Bumiputera titular category and through that category is represented. In this particular context of mixed constituencies with Muslim plurality, and with the Kedayan being dominant, Bekenu and Lambir become "Kedayan seats". Lambir is also an exception to the iron rule of PBB contesting Muslim seats. Lambir was allocated to SPDP and the Kedayan Rosey Yunus is the only Muslim fielded by a party other than PBB within BN. Therefore, because of sophisticated gerrymandering, the Kedayan found themselves in a position to be entitled to a Kedayan representative (*The Borneo Post* 2006b, 2010b), although in any other context they are likely to be seen simply as Muslim Bumiputera.

Kedayan is also the biggest of the Muslim categories in the parliamentary seat of Limbang (mixed constituency with no clear majority). The elected representative for the Limbang constituency is a Malay, and fielding him – and not a Kedayan – as a candidate almost cost PBB a defeat in the seat in 2008.[12] On a constituency level, therefore, "Muslim Bumiputera" is not a minimum-winning coalition in its own terms, or in other words, voters have a preference as to which particular category within the Muslim Bumiputera is fielded in their constituencies. The linguistic category "Kedayan", also the only pocket of Muslims dwelling in the interior, remains activated here, although it is not a titular category in Sarawak. Similarly, Lun Bawang is a category that by itself constitutes a minimum-winning coalition in the sole seat of Ba'kelalan. Although the constituency is classified as an Orang Ulu seat, it is the Lun Bawang who are a majority in the seat and it is Lun Bawang candidates who contest against each other in this seat (compare Puyok 2005 and see later in this chapter). Therefore, on the level of constituency there is a search for a localized minimum-winning coalition and the constituency level is the most sensitive one to re-delineation exercises. The ethnic composition of Lambir, Bekenu and similar constituencies can be changed by moving a mere few hundred people from one constituency to another. This way, an ethnic category can be upgraded to a plurality that will entitle them to claim the seat and subsequently have a candidate from their local minimum-winning category contest the election.

To bring this point home, note that the activation of the Orang Ulu category happened through designation of constituencies in Sarawak as "Orang Ulu". The majority of voters in these constituencies belong to several linguistic categories (Kayan, Kenyah, Kelabit, Lun Bawang and several others). Each of these constituencies, both assembly and parliamentary seats, consists of a mix of these linguistics groups, but for political purposes they are mobilized as "Orang Ulu" and their share in power is realized through the category "Orang Ulu" and not "Kayan", "Kelabit" etc. This is not to say that these linguistic categories are of no political use. When BN fielded the first Bisaya (Orang Ulu) in the 2006 state election, his candidacy announcement included information on his Bisaya background (*The Borneo Post* 2006a). However, although the local minimum-winning coalition may be vested in a local majority category ("Kedayan", "Lun Bawang", "Selako Bidayuh"), the title to share the power is rooted in the titular category in which the local category can claim membership. A similar process happened in the Bidayuh areas; instead of perceiving themselves as members of ethnic categories divided by languages (Bukar, Sadong, Jagoi or Biatah), their political representation happens through the Bidayuh category. As Michael Jagoi, a Bidayuh state representative, emphasized, "I don't see Bukar, I don't see Sadong. I only look at Bidayuhs in unity" (*The Borneo Post* 2006f). The different linguistic groups are a majority in six state seats, which is their share of power. Moreover, the four linguistic categories are entitled to participate in the power bargain through the Bidayuh categorization of their constituencies.

Numerous examples, however, show that the fixedness of constituencies mentioned earlier does not always lead to the exact same ethnic composition of the assembly after each election. Most prominent cases that defy the static and automatic process of ethnic nomination are those of non-Muslim indigenous seats and their winning Chinese candidates. The 1980s case of SUPP Yong Kuet Tze in a Bidayuh seat and James Wong Kim Min in Limbang (a mixed seat) are the two most prominent cases. More recently, it is the Iban majority seat of Pelagus in the heart of Rajang Iban territories that had Chinese candidates elected over four consecutive elections. Although the patronage element of these electoral successes is strikingly obvious, the last case will be studied in more detail to observe the mechanisms of ethnic marketing and mobilization that accompany this type of ethno-clientelistic process. All four non-Muslim indigenous parties (SNAP, PBDS, PRS and SPDP, the successor of SNAP) experienced severe internal power struggles at some point of their history resulting from discontent over what was perceived as excessive Chinese influence over the party. The financial aspect of the phenomenon is only too obvious. SNAP was financed first by James Wong Kim Min, later by Tiong King Sing. Sng Chee Hua's wealth was an enormous asset to PBDS and later to PRS (James Masing, interview, *The Borneo Post* 2006e). Also voters accept the Chinese candidates from the non-Muslim indigenous parties readily and vote for them as representatives.

From 1991 Sng Chee Hwa, a local wealthy Chinese from PBDS,[13] was Pelagus' elected representative in the state assembly. Sng's nomination was only possible thanks to the recent decision of the party to relax its Dayak-only doctrine of membership. The Pelagus (Iban) seat was inherited in 2001 by Sng's son Larry (21 years old at the time) who won easily in his father's constituency, also on the PBDS ticket. What ethnic logic is there to a Chinese being not only fielded but also winning a non-Chinese constituency? In the case of the Sngs in Pelagus, it may be attributed to their membership in PBDS – the most outspoken Dayak party Sarawak has ever had. This fact would support the assumption that ethnic representation materializes not necessarily through the ethnic identity of the candidate himself, but the ethnic orientation of his party. This might have been the case when the senior Sng successfully contested Pelagus against a local Iban, grandson of Temenggong Jugah, Alexander Nanta Linggi, in 1991. Alexander is from PBB (the Pesaka wing), which is associated with Malay/Melanau interests, despite its Pesaka non-Muslim wing. After the deregistration of PBDS, Larry continued to be a member of PRS and competed against James Masing for the chairmanship of PRS. The power struggle in PRS between Sng and Masing ended with Masing's victory and Larry Sng was expelled from the party. However, Larry Sng remained, despite being party-less, on good terms with the chief minister; he was re-nominated as a member of the cabinet in the rank of assistant minister with three portfolios (*The New Sarawak Tribune* 2010).

In the light of all the assumptions about power-sharing arrangements, which tell us that party (in this case a non-Muslim indigenous) and constituency majority (here Iban) are carriers of ethnic activation, one has to ask whose representative Larry Sng was (or his father before him)? Did he represent the Chinese or the Iban? The chief minister, when asked why he was willing to retain in the cabinet a person who clearly did not have the support of his own party, replied that "I would like to see Sng retained because we do not have enough young Chinese leaders" (*Bernama* 2009). This way Taib indicated that Sng is not only the representative of his Iban constituents, but also a Chinese leader.

Because of the complicated status of the incumbent Larry Sng, his former party, Parti Rakyat Sarawak, was in trouble before the 2011 state election (*The Sunday Post* 2010). As a non-member, Larry could not be nominated by PRS. James Masing, the PRS president, chose an Iban, Stanley Nyitar, to contest in the seat, arguing that "the people in Pelagus had requested for a local born to stand in the constituency. I gave in to their request by nominating their own people in the person of Stanley Nyitar who was from [the] Merit area" (*The Borneo Post* 2011d). Larry Sng did not contest the election, but an independent, George Lagong, also contested in the constituency and in the end won. "Coincidentally, Lagong who beat BN candidate Stanley Nyitar by a 2,837 vote-margin is [Larry] Sng's father's half-brother" (*The Borneo Post* 2011c). The constituents, as it turns out, did not mind the Chineseness of the Sng family and voted for its proxy in the election. In fact, there is a long history of Chinese candidates contesting and winning non-Muslim indigenous seats. The first case was the then-Iban state seat of Igan (currently a parliamentary seat with a Melanau majority):

> S[tate seat] 29, Igan, was won by the SCA with a Chinese candidate, although the seat had more Dayak than Chinese electors. The SCA candidate, however, Dato Ling Beng Siong, had unusual qualifications. He had cultivated the constituency for a long time, had made personal donations in the area on a truly grand scale, spoke fluent Iban, and had a Dayak wife.
>
> (Milne and Ratnam 1972, 119)

Needless to say, patronage-related calculations suggest a parallel explanation of the voters' willingness to accept representatives with whom they share membership in no ethnic category. What is the mechanism, should we ask, behind voters supporting candidates who are from a different ethnic category? As was stated, the (titular) ethnic homogeneity of voters and candidates in a given constituency is the institutional default. It is a basic right in the Sarawakian political bargain: voters vote for "their own" people who then represent their interests. Therefore, there is no particular leverage against a representative who shares ethnic background with her constituents. However, let us take Iban constituents in the Kidurong seat (about equal numbers of Ibans and Chinese) and ask what candidate the Ibans should prefer. Arguably, they are likely to have a stronger leverage over a Chinese representative than they would have over an Iban candidate. It is in the Ibans' best interest that the Chinese candidate continues to court them for support; an Iban candidate would be more concerned with the expected poor support of the Chinese electorate and would be busy vying for votes among the Chinese, almost a lost cause in the current political climate in Sarawak. No surprise, therefore, that main candidates in Kidurong, a mixed seat, have always been Chinese, whether from BN or the opposition.

Moreover, Kidurong is an excellent case to be presented as an answer to the obviously striking question: in the case of a mixed seat of an almost even split, would the opposition candidate likely be from the alternative ethnic category, i.e. a different one from the incumbent? To put it in different terms: does the opposition aim to mobilize "the other category" to gain leverage over the ruling coalition? The Kidurong seat includes the town of Bintulu, which has an overly Chinese population, but the rural area around the town is majority Iban, and the Iban comprise almost half of the constituency (*The Star* 2011a). The Democratic Action Party (DAP) won a 1996 Chinese-against-Chinese contest in Kidurong, and repeated this performance in all consecutive elections since (*The Borneo Post* 2011a; *The Star* 2005). After the 2006 loss by the SUPP candidate, SPDP and PRS both requested that Barisan Nasional revise the party allocation for the seat, quoting an alleged internal BN rule that after a component party loses in a constituency twice, other components should be given a chance to contest there (*The Borneo Post* 2009). Whether SPDP or PRS would have fielded a non-Chinese candidate remains unknown because there was no revision and DAP safely won the seat against SUPP with a Chinese candidate.

DAP, perceived in the vernacular as a multiethnic party with strong Chinese inclinations, had an obvious incentive to field a Chinese candidate in the seat; nevertheless, if the party sensed a whiff of potential amongst the Iban to turn against their Chinese representative, it might have fielded an Iban candidate. It did not; as pointed out earlier, inter-ethnic competition is avoided at all cost, and in fact within the young opposition coalition Pakatan Rakyat seats are often allocated based on the BN allocation. If SUPP contests for BN, DAP will be assigned the same seat within Pakatan Rakyat, a rule rooted not only in the ethnic majority paradigm in seat allocation, but also in party competition between the two coalitions. So went the logic of the DAP candidate in the Iban-majority Sri Aman constituency in 2011, contested by SUPP on the BN side.[14] Both SUPP and DAP fielded an Iban there; in the end, SUPP retained the seat.

Therefore, candidates from outside the minimum-winning coalition category are not unwelcomed by the voters. The candidacy's legitimacy in such cases does not rely on the ethnic identity of the candidate, but on his merit, expressed in patronage terms. The Sng case sheds some light on this: Larry as a rising star of the chief minister's administration was probably the most "valuable" representative the seat could have in any case, and the fact that he was a Chinese would keep him on the tip of his toes because any Iban candidate would have a stronger claim to the seat than him. In case of candidate-constituents ethnic discrepancy, the voters automatically exercise some influence on their assemblymen, which otherwise – in the case of candidates by ethnic default – has to be exercised through a sophisticated and difficult to execute "lowest winning margin" scheme, presented by Faisal Hazis.[15]

The most important finding for this research is the fact that voters' behaviour is ethnically motivated even though they support a candidate who shares no ethnic category with them. Although one might argue that Iban voting for a Chinese candidate are "non-Muslims voting for a non-Muslim", there is little evidence to support such an interpretation, although it is probably not far-fetched. I argue, however, that the Iban are acting in their best, also ethnically defined, interest when they allow and seek for their constituency to be represented by a candidate who does not share an ethnic membership with them, but whose political fate entirely depends on their support.

In the previous paragraphs, it was shown that political practice often deviates from the rules of power sharing in Sarawak. The deviation, however, does not affect BN's popularity and legitimization. Voters welcome the potential additional leverage they gain from having as their representative someone who is not from the majority category.

3.5 Opposition and ethnic mobilization

Any opposition to BN, or to any permanent multiethnic coalition for that matter, can follow three strategies: one is to focus on the fringes of an ethnic category and aim at ethnic outbidding (Rabushka and Shepsle 2009). Another strategy is to offer an idea of non-ethnic or de-ethnicized politics in place of the existing ethnic bargain. Yet another option is to take over power by copying the BN model and creating an alternative but mirror in outlook coalition. Sarawakian contenders against BN were historically of all three types.

SNAP in opposition (1966–1976) was shown to shift gradually towards non-ethnic opposition, and had it not joined BN, it would likely become a non-ethnic party. PBDS and PERMAS as opposition parties were clearly ethnic, and PBDS with its Dayak ethnonationalism followed the outbidding strategy (SNAP was the casualty; see the previous chapter). The current Pakatan Rakyat (People's Pact or PR) coalition, established as a permanent one in 2008, is in many ways a variation of BN, especially of the Sarawakian chapter. Except for the Islamic party Pan-Malaysian Islamic Party (PAS), the other components of PR are not ethnically exclusive; on the contrary, both claim to be multiethnic. The multiethnic slogans, however, seem to be directed mostly towards the non-Muslim indigenous in the case of both DAP and Parti Keadilan Rakyat (People's Justice Party or PKR), much like SUPP and PBB in BN. Both respective parties are associated with a particular category: DAP with Chinese, and PKR with Malays (less so, however, than PAS), and the main question for the Pakatan Rakyat in Sarawak is: who should represent the non-Muslim indigenous? Later I discuss this issue, which transpired during the 2011 Sarawak state election.

Since 2010 SNAP, resuscitated through a court decision after being unlawfully deregistered, had been a member of the Pakatan Rakyat coalition, originally a West Malaysian cooperation of three parties: DAP, PAS and PKR. Prior to SNAP's entry, the assumed seat split would have been: DAP – urban Chinese, PAS – high Muslim majority, PKR – others. None of the three parties had a strong claim to non-Muslim indigenous seats. DAP has had several elected state assemblymen over the years, all of them urban Chinese–majority seats. Although DAP attempted for years to extend its support to rural areas, it never succeeded. Arguably, PKR wished to establish itself as the centre party and the most genuinely multiethnic one within the PR, and therefore, the most suitable to capture the non-Muslim indigenous vote. However, PKR had won only one seat in Sarawak before: in 2006 Dominic Ng, a Chinese, won the Padungan seat in Kuching. PAS was new to the state and SNAP, although a well-known brand on the political scene, was yet to prove that the re-registered party carries any of the values and strength of the old SNAP.

Therefore, in 2011 there was no credible information about voters' preferences towards particular opposition parties. The seat allocation was done based on quite elusive criteria, e.g. which party claimed to be more popular among the grassroots, or which party had a more winnable candidate for the seat. According to Tian Chua, the PKR West Malaysian leader, PKR's criteria for candidates were: NGO background, community service and no prior involvement with BN. Since 2009 Baru Bian, the lawyer *cum* preacher,[16] a Lun Bawang Orang Ulu, was the state liaison to the national PKR leadership and within Sarawak, he had the authority to recommend candidates for the state election.[17] Final negotiations were led by Mohd Azmin Ali, the deputy president of PKR; PKR contested 49 seats, DAP 15 seats, and PAS 5 seats.

SNAP contested only non-Muslim indigenous seats and had no claims to constituencies assigned to DAP or PAS, but found itself at odds with PKR over these seats, resulting in SNAP's exit from the coalition. In the words of Tian Chua, "PKR is trying to use class identity to overcome ethnic identity",[18] although this line of thinking, admitted Tian Chua, has not yet been incorporated into an official party manifesto. According to Tian Chua, although PKR was "trying to undo the ethnic division", in practical terms the party is still operating in an environment where ethnicity is perceived as the title to share power – whether within BN or PR. Although Tian Chua and See Chee How insisted[19] that PKR does not deploy ethnic background as a criterion for selecting candidates, a study of their candidate list actually suggests that PKR nominations followed the ethnic path for nominations in terms of ethnicity (*The Star* 2011b). Moreover, given that DAP contests the Chinese seats, and PAS (despite its negligible popularity in the state) has to be allocated some of the Malay or Melanau seats, PKR's best hopes are in the non-Muslim indigenous seats. These are easier to win than the Muslim ones, and coalition partners (except for SNAP) have no particular ethnic claim to them. SNAP's decision to quit Pakatan Rakyat was a result of disagreement over seat allocation with PKR (*Malaysiakini* 2011). SNAP and PKR both eyed non-Muslim indigenous seats.

DAP has a long tradition of Chinese support, and bears a strong resemblance to the SUPP in terms of its ideology and its ethnic fate: a party that wished to be for all, but is seen as Chinese in the end, especially within a coalition that pushes it to represent a particular section of the society. DAP traditionally contests urban Chinese seats, although in its rallies and promotional materials, it would be difficult to find statements indicating the party's Chinese appeal – except for the fact that most of the DAP statements are made in Mandarin. In ethnic terms, a Chinese DAP MP from Kuala Lumpur compared Muslim domination in Malaysian politics to white domination in American politics before Martin Luther King. The MP, after recalling Rosa Parks' story and fast-forwarding to Obama's wins in presidential elections, went on to ask when a non-Muslim would be able to become the prime minister of Malaysia.[20] This comparison would have resonated with the non-Muslim indigenous voters in Sarawak even better than with the Chinese, but because the MP spoke Mandarin, it is unlikely that the message reached significant numbers of non-Chinese voters.[21]

Prior to the 2011 state election in Sarawak Pakatan Rakyat also wished to send a strong message pertaining to the leadership in the coalition, in order to strengthen the coalition's credibility and prove its worthiness as the BN contender. Baru Bian was unofficially proposed to take up the chief minister position in case PR won a majority in the Sarawak state assembly (*The Malaysian Insider* 2011). As the state chief of the biggest coalition partner at the national level, he would be the natural candidate for this position. The difficulty with this decision was twofold. For one, PKR would have to win more seats than DAP to claim the chief minister's position – it is, after all, the most successful party in a coalition that nominates the chief minister.[22] At the time when this decision was publicized, however, it was DAP and its Chinese candidates that were PR's best hopes and Baru Bian could not even be sure of his win in his own constituency. It was therefore easy to see through these marketing operations aimed at showing the strong non-Chinese face of the opposition, but that were inconsistent with the parties' electoral strength within the state.

Therefore, the main dilemma presents itself not when selecting legislative candidates, who are, after all, always of the same background as the majority of a constituency, but on the occasion of selecting state-level leaders. Since 2009 in Sarawak Baru Bian has served as the state chief of PKR. Jimmy Donald's (in 2011 PKR, formerly PBDS/PRS) statements may suggest that the popular feelings, shaped by decades of power assigned according to numerical proportions, are following the old path of thinking:

JD: For that reason, I think, I am one of those, who think that Baru [Bian] cannot be a leader of PKR in Sarawak.

KP: *So who could be?*

JD: An Iban, or a Malay.

KP: *Either Iban or a Malay?*

JD: Yes.

KP: *Why do you exclude the Bidayuh etc. . . . ?*

JD: We are race blind, but we have to be race sensitive. It is unthinkable. There is only one Lun Bawang seat in Sarawak. That is the seat that Baru is standing on. I cannot agree [to have Baru Bian as the leader of PKR], because there is no sense of belonging. You must know the Lun Bawang mentality. They go out of their way to say that they are not Dayak. They go out of their way to make themselves different from the other Dayak. [. . .]. So if you go out of your way to tell me that you are different, but the next day you tell you want to lead me, I say, I don't feel that I belong to you.

KP: *But Malays are even more different from Dayaks than Lun Bawang [aren't they]?*

JD: Yes, but they have something to give! There are 23–24 Iban seats, 6 Bidayuh seats, 3 Orang Ulu seats, about 15 Chinese seats, about 19 Malay seats. So politics is an art of using and being used. So if you are a Malay leader you need me, but I can give you 19 Malay seats and together we can form the government.

KP: *What you are suggesting is ethnically based thinking.*

JD: Yes, but I said, ethnic based, it must be fair. I can accept a Malay leader, a Chinese leader, as long as he's fair.

KP: *But not a Lun Bawang leader.*

JD: No, because he cannot be fair!

KP: So you are saying the Dayaks here could accept a Malay leader as long as he is against the Malay supremacy? Could the Dayaks here see Anwar [Ibrahim] as a national leader?

JD: I think we could; I have no problem accepting a Malay leader, as long as he's fair. I think I have no problem. I would be very comfortable with that.

KP: How about a Chinese leader?

JD: I would be very comfortable too. No problems there. As long as he's fair.
(Interview by the author, 14 December 2010)

Jimmy Donald's opinion is quoted here not to attach any weight to his influence in the party or in Sarawakian politics, which was very limited at the time of the interview, and Jimmy was later expelled from the party's leadership for undermining Baru Bian's position. Although this is clearly not PKR's official message, it is likely that some voters would indeed feel this way. Iban is the category within non-Muslim indigenous that not only by the token of numbers, but also by the token of tradition, perceives itself as the one that should be in charge of Sarawak. To substantiate the point, let us recall the earlier mentioned situation in SNAP in 2002, when Orang Ulu were discarded as potential leaders. Edmund Langgu, an Iban and former leader of PBDS, also stated that Orang Ulu are not suitable as leaders:

> He [Baru Bian]'s just the leader for Sarawak [but not at the national level]. Unfortunately Baru Bian came from a small community. It would be different. If I become a leader, it would be different. I am not overclaiming, but [. . .] I'm not bragging. If I am a leader, it would be different, because I am coming from a bigger community [Iban], also probably because of my previous influence, I got more ground [. . .] but for my friend Baru Bian, he's still quite green in politics. And also, unfortunately he came from a very small community, from [Lun Bawang].
> (Interview by the author, 30 September 2010)

The results of the 2011 elections in the end tell nothing about the Ibans' preference for a native Christian non-Iban chief minister or party leader. It seems unlikely that the relatively poor showing of PKR (three seats won, a Chinese, an Iban, and a Lun Bawang/Orang Ulu) is to be associated with the ethnic background of the party's leader; note that PKR's nationwide leadership is devoid of any non-Muslim indigenous Sarawakians; to the contrary, the party's leaders are mostly West Malaysian Malays, a category that poorly resonates in Sarawak.

The results of the election were as expected: both DAP and PKR improved compared to 2006; DAP had 12 representatives (all Chinese), PKR had three. DAP's sweeping victory overshadowed PKR's result and although Baru Bian (PKR's chief in the state) became an assemblyman, it was DAP's Wong Ho Leng who became the opposition leader in the assembly. Therefore, although the opposition wishes to send a message that it would return power to the non-Muslim natives in the state (Baru Bian's role suggests this much), the political reality may be that it is the Chinese who will grow most politically powerful in the opposition.

3.6 Ethnic identity change in Sarawak: conclusions

At the time of fully competitive elections and in the absence of any particular political tradition to follow, the 1963 (indirect) and 1969/1970 elections were a good measure of how the West Malaysian power sharing could be reproduced in Sarawak only with great difficulty. Ethnic identities were fluid, parties' electoral bases even more so; moreover "the absence of a distinct politically dominant community, coupled with the fact that more than one party claimed to represent some major ethnic groups, meant that ethnic considerations could not provide decisive guidelines for the allocation of seats [within the Alliance]" (Milne and Ratnam 1972, 152). The same problem of fluidity and lack of majority was reflected in the search for an agreeable chief minister. In the absence of UMNO-style backbone – as the hierarchically highest party – there was no clear indication as to who, or at least from which party/ethnic category, should be the chief minister. Upon the establishment of PBB the problem seemed solved, but without the self-explanatory certainty UMNO enjoyed. PBB retains the position of chief minister rather by the virtue of inertia and support from Kuala Lumpur than by renewed mandate from its coalition partners.

Between 1963 and 1976, i.e. prior to SNAP's entrance to the cabinet, parties were ethnic as part of the ruling coalition, but non-ethnic in opposition. This was the case of SUPP, which went to great lengths to maintain its image as multiethnic and leftist when in opposition, but turned pronouncedly Chinese and definitely non-programmatic when in the government. With SUPP in the government, SNAP remained the only relevant opposition party between 1970–1976 and proved able to attract all the protest vote from all communities (although less so from among the Muslim indigenous). As it was shown in the example of several Sarawakian parties, being part of the multiethnic coalition proscribes ethno-exclusive mobilization and makes the coalition members vulnerable to tactics of ethnic outbidding by representatives of the respective ethnic categories outside of the coalition.[23] So was the case of DAP, Permas and PBDS when these parties were competing on the state level against the ruling coalition.

The striking point is that for extended periods of time both Rahman Yakub and Taib Mahmud as chief ministers managed to maintain ethnically all-inclusive cabinets, which, however, did not include parties representing some categories that were relevant at the time. This was specifically enabled by the fact that Sarawak did not strictly follow the West Malaysian model of one community corresponding to one party. Having multiethnic parties allowed for coalitions that in their outlook indeed comprised all relevant ethnic categories, but excluded parties that would be a threat to the position of the Muslim component party or would openly demand to nominate the chief minister, i.e. SNAP until 1976 and PBDS 1983–1994.

Specifically for the last reason Chinese-led parties were convenient partners for the Muslim party. Both SCA and SUPP were well aware that the deputy chief minister is as high a position as they will ever get, and they would not contest the top position in the state. On the contrary, installing a chief minister that would be an accommodating partner for the Chinese was much more important for SCA and SUPP than promoting a Chinese chief minister who could never be accepted in the greater Malaysian context. This is also the answer to the question of why the Bumiputera category never really materialized in Sarawak (except in the period 1966–1970, but even then it was marriage of convenience). The decades of coalition negotiations and renegotiations show that it was the Chinese component that has been the party to guarantee the necessary majority in the parliament, while the non-Muslim indigenous components keep joining the coalition as the additional and last in terms of precedence, and therefore disposable and weak component. So was the case of Pesaka in 1970, SNAP in 1976 and PBDS in 1994. The multi-ethnic image of the cabinet was easily achieved by token ministers supplied either by the Pesaka component of PBB or non-Chinese SUPP elected representatives. The most prominent case was the first Rahman Yakub cabinet, whose two Iban ministers were coerced to join (1970), while the Bidayuh minister was instrumental to win a by-election (see the previous chapter).

"Dayak" as a category was assigned a political value in the early 1980s and from this time on it is crucial to distinguish between "Dayak" and "Iban", "Bidayuh" and "Orang Ulu" as activated categories. All four have been activated in political life, but clearly have very different political targets. The lack of success in activating the "Dayak" category as the minimum-winning coalition in itself can be traced back to the supra-institutional BN arrangement. "Dayak", although numerically strong and possibly a convenient coalition partner for the Chinese (judging by inner-party cooperation), can hardly hope that the UMNO/BN president *cum* prime minister would agree to have a non-Muslim chief minister within BN. The guarantee that a Muslim would always be the chief minister has two implications. Firstly, it gives PBB the advantage of being an (or better put, *the*) indispensable coalition partner. Secondly, consecutive Malay/Melanau-dominated cabinets simply helped to increase the representation of Malays and Melanaus by creating new Muslim-majority constituencies and, in turn, legitimizing a Muslim chief minister in the long run.

From 1994 (when PBDS joined BN) onwards, inter-ethnic competition in elections has been minimal. Nevertheless, ethnic manoeuvring remains intense. Political parties keep reinventing themselves as "Dayak", "rural", "multiethnic" etc. Therefore, we have "Dayak-based" parties, but "Iban" or "Bidayuh" ministers. Moreover, "Dayak-based" parties are making a conscious attempt to part with the "Dayak" image, which indeed they merely inherited from their predecessors, and try to re-establish themselves as "rural parties" (PRS) or "multi-racial parties" (SPDP). This is visible either in their new modern logos (see Appendix 1 for PRS) or in their leaders' statements. At the same time, parties have to maintain a careful balance when allocating ministerial seats to non-Muslim indigenous categories not to cause resentment among the Iban, Bidayuh and Orang Ulu.

The indispensability of PBB for the cabinet is a strong message for both the Chinese and non-Muslim indigenous elites, who, knowing the federal preference for a Muslim executive, are more inclined to seek close cooperation with a Muslim chief minister hoping for the number two position, than to challenge the top position directly. One example of this inclination was the 1987 election; the leader of the Dayak component of the Maju coalition (PBDS and PERMAS) did not contest the state election, making clear that it would be the Muslim leader of the coalition to take the chief minister seat. The 1987 election was, after all, an internal BN coup attempt and the contending coalition was well aware of the BN rules; in the end their survival as a government would have depended on support from Kuala Lumpur.

Strikingly, given the distribution of seats between the ethnic categories in 1987 (compare Table 2.3), non-Muslim indigenous was then a minimum-winning coalition on its own, with 25 out of 48 seats being Iban-, Bidayuh- or Orang Ulu–majority constituencies. But the non-Muslim indigenous category has politically little to offer as long as BN exists; although it can claim numbers, it cannot take over the executive. Would the arrangement be any different under the rule of the current challengers, Pakatan Rakyat? There is little information to go by, as leaders' statements are hardly reliable; however, Pakatan Rakyat is as much a product of West Malaysian politics as BN, and follows similar guidelines for candidate selection as BN. In fact, to notice that UMNO opened up to non-Muslim Bumiputera in Sabah and can be now called a "Bumiputera" party suggests that it is the equivalent of PKR in Pakatan Rakyat (although UMNO is highly unlikely to ever accept Chinese members, unless they convert).

However, if it were as simple as that, one would have to conclude (and some do), that the Muslims entirely Sarawakian politics, which would imply that the support of other categories is not necessary for the stability of the government. This is far from the truth in Sarawak: support of all categories is equally sought after, and not only for legitimization. The Malay and Melanau seats cannot be multiplied to the extent to produce a minimum-winning coalition on its own; the geography of ethnic distribution in Sarawak precludes that. Muslims can ill afford to antagonize other categories, as each seat won for the federal parliament is an extra merit point for the state government in the eyes of the federal BN. Therefore, the Sarawak government and parties within it resort to most inventive manoeuvres to make sure that in each constituency there is a candidate whose ethnic background produces a minimum-winning ethnic category in terms of the ethnic background of the constituents.

This way we arrive at by far the most important finding of this study: a power-sharing scheme does not have to produce fixed ethnic categories. This finding is specific for the case of Sarawak, which is characterized by undefined conditions of the power-sharing arrangement. The power is shared through multiple channels: 1) parties, 2) constituencies and their majorities, 3) ministerial posts on both the state and federal levels. In fact, even distribution of Senate nominations and BN component parties' executive positions are a matter of scrutiny in terms of ethnic strength. Thanks to these multiple dimensions on which power is shared, various categories are activated for each channel and remain activated over dozens of years. The power-sharing scheme in Sarawak is based on a margin of flexibility of ethnic mobilization.

This flexibility enables the constant activation and de-activation of categories. Most importantly, parties cannot be precisely associated with any particular category, consequently, neither can their elected representatives, and among them, ministers. This way, each party and candidate must be ready to identify with any of the categories in which they can claim membership, e.g. "Bumiputera", "Dayak", "Bidayuh" and "Selako Bidayuh" are all equally likely to be activated at some point of the political process. Chinese candidates also have a choice to either speak Mandarin, and by this token activate their "Chinese" identity, or to speak a dialect, by which they will be seen as a Foochow, Hakka or Teochew. If needed, a Chinese assemblyman elected in an Iban seat can be seen as a champion of his Iban constituents' interests, or a promising leader of Chinese Sarawakians. A party can be a Dayak party, a multi-racial party, a rural-based party or a "Sarawak new generation" party. The bottom line is, keeping all options open is the rule of the game. Sarawakian politicians carefully maintain membership in each category; each of them is deployed with certain frequency to make sure that none becomes too prominent. Exclusiveness is the most dangerous accusation in the Sarawakian context. Therefore, constant manoeuvring of identities does not have to originate from multiplicity of elections on several tiers. Anchoring the power sharing in multiple channels (parties, seats, executive and senatorial nominations) enables ethnic identity change equally well.

Further studies of consociational political systems must be conducted to test discoveries of this research. Here it was revealed that particular institutions played their own roles in the power-sharing scheme, or, better, the institutions were deployed to facilitate the power sharing. The first-past-the-post electoral system with single-mandate constituencies have two important consequences: constituency size is relatively small and even relatively small categories (Melanau, Bidayuh, Orang Ulu) become titular categories in some of them, and in turn become units entitled to share some political power. In bigger constituencies with multiple mandates these three might cease to constitute a minimum-winning coalition in any constituency (depending on size and delineation), and render them irrelevant.

To conclude, Sarawak is an example of a consociational polity in which at least two sets of ethnic identities are continuously activated in the society. One set splits Sarawakians into three categories, while the other splits them into five or six categories. Each set is activated in a different context: the first through party politics, the second through ministerial and senatorial nominations as well as legislative candidates. Therefore, we see that consociational designs do not preclude maintaining of more than one ethnic identity in each person's repertoire, and shifts between these categories can indeed be very frequent. Note, however, that no difference was found in activated categories on different tiers of legislative elections. The same set of categories was activated in state and general elections.

The temporal changes of ethnic identity were shown to be related to institutional development. Different sets of identities were activated prior to 1969; the West Malaysia–inspired political parties and party system induced the first change of identities on the time axis. Later mobilization of the Bidayuh and Orang Ulu categories through intense explicit electoral campaigning in these areas activated these two identities. Creation of new constituencies in which lesser categories constitute a minimum-winning coalition is the only way in the current political reality of Sarawak in which new categories can be activated in the future. Finally, several categories, albeit existent in Sarawak, are not activated in politics; most important among these are "Bumiputera", "Muslim" and "Christian". Although these categories come up in public discourse, they are not platforms of political mobilization.

To relate to the theory of coalitions, it was shown that BN as a meta-structure renders a coalition composed only of Chinese and non-Muslims indigenous merely as a blocking coalition. Because it would not include the Malays and Melanaus, it could not be a winning one. Activation of the "Dayak" category could be more successful under a proportional electoral system, as will be shown in the case of Indonesia. A "Chinese + non-Muslim indigenous" coalition could be a winning one in the case of direct executive elections.

Notes

1 To illustrate this point: after the 2005 delineation exercise only 11 out of 71 state seats were part of parliamentary constituencies that were of a different ethnic majority. This refers to titular categories ("Orang Ulu" is a current titular category for constituencies; "Kedayan" and "Penan" are not). In early 2015 the Election Commission embarked on the next re-delineation process; there were 11 seats to be added. However, at the time of writing, the ethnic composition of the proposed new seats was not yet known.
2 Rosey Yunus, a Muslim Kedayan female candidate from SPDP in the Muslim plurality seat of Bekenu (2011 state election), is an exception.
3 Chin (1996) presents an overview of each election SUPP contested until the early 1990s, with a brief account of each seat contested.
4 The sole lost SUPP Iban seat (Engkilili) was contested and won by a "BN-friendly independent", Johnical Rayong, who stood against an official SUPP candidate in Eng-kilili, and subsequently applied to join SUPP; his admission took place in 2010, when the party had to select its candidate for the seat in the upcoming state election (*The Borneo Post* 2010a).
5 Keep in mind that in the light of the Sedition Act, even internal BN or parliamentary debates are forbidden and championing Chinese interests would border on sedition in the Malaysian context.
6 I owe this information to Ngu Ik Tien (e-mail communication, 2 February 2012).
7 Actually, Wong Soon Koh "boycotted" the election, along with several elected representatives from the party (*The Star* 2011c).
8 In the 2008 general election SUPP won six seats, five Chinese and one Bidayuh (Richard Riot). In 2011 the Chinese parliamentary seat of Sibu was lost to DAP in a by-election. In 2011 only two Chinese SUPP candidates won their seats to the state assembly.
9 All the "presidents" posts are uncontested (*The Borneo Post* 2006d); the line-up of senior party positions is usually agreed via "compromise" prior to the triennial general assembly that each party has to hold. All four BN parties submit to this practice.
10 The 2007–2009 PRS dispute between the Sng faction and the Masing faction is very informative; this party, however, survived the turbulence. The 2002 SNAP deregistration followed a similar scenario (*Malaysiakini* 2002b). At the risk of stating the obvious, only the "Dayak" parties (and recently the SUPP) suffer from the notorious problem of having multiple presidents and being presented a show-cause letter by the ROS.
11 Interestingly, this was precisely MCA's situation in West Malaysia after the 1969 election; then "MCA withdrew from the Cabinet on the grounds that it had lost the confidence of Malaysia's Chinese community" (Wicks 1971, 19). This move was, however, of no consequence as a state of national emergency was introduced after the 13 May

riots, the National Operation Council took power (Kua 2007, 49–50) and a new, grand coalition arrangement followed in 1970.

12 I owe this information to Faisal Hazis (personal communication, 21 July 2010).

13 The extent of the Sngs' business connections is enormous; suffice it to say, in 2006 Larry Sng married May Ting (*The Star* 2006), the oldest daughter of Ting Pek Khiing, who is one of the biggest Sarawakian tycoons and an immediate business partner of Taib Mahmud, the Sarawak chief minister of more than 30 years and the current governor in the state (Gomez and Jomo 1999, 110).

14 Tian Chua, interview by the author, 14 April 2011

15 Faisal Hazis (2012) found that voters want to keep their representative on the tip of his toes so that he continues to court his constituents. If he wins comfortably, he becomes complacent, so goes the logic of the voters. In order to avoid the representative's complacency, the voters try to hold the electoral win within as narrow a margin as possible.

16 Baru Bian is a popular figure of the Sidang Injil Borneo (Borneo Evangelical Church, SIB) and a successful lawyer who specializes in Native Customary Rights court cases; he has represented indigenous communities when they press charges against the government for unlawful land acquisition (Baru Bian, Native Customary Rights talk, held in Kuala Lumpur, 1 May 2010).

17 Tian Chua, interview by the author, 14 April, 2011.

18 Tian Chua, interview by the author, 14 April, 2011.

19 Interviews by the author, 14 April, 2011 and 5 August, 2010, respectively.

20 Teo Nie Ching's speech at the DAP rally in Kuching/Batu Kawa on 13 April 2011.

21 Although, according to Ngu Ik Tien, about 20% of students in Chinese-medium schools in Sarawak are "native" (Ngu 2011, 11).

22 The mathematics of PR's potential win was complicated: DAP contested 15 seats, PAS 5, and therefore, in order to capture a simple majority in the assembly, PKR would need to win 16 of its 49 seats, with the coalition partners having a 100% success rate. Therefore, a situation in which PKR wins fewer seats than DAP is equivalent with BN winning the election altogether (unless we consider the highly unlikely situations of multiple independents or SNAP candidates winning their seats and subsequently joining PR).

23 Textbook cases are UMNO and PAS in West Malaysia.

References

Bernama. 2009 "Six new faces appointed assistant ministers in Sarawak". *Bernama Online*. www.bernama.com.

Chin, James. 1996. *Chinese Politics in Sarawak: A Study of the Sarawak United People's Party*. South-East Asian Social Science Monographs. Kuala Lumpur, Malaysia and New York: Oxford University Press.

Free Malaysia Today. 2011. "Bidayuh lose out to Orang Ulu again". *Free Malaysia Today Online*. www.freemalaysiatoday.com.

Gomez, Edmund Terence, and S. K. Jomo. 1999. *Malaysia's Political Economy: Politics, Patronage and Profits*. Cambridge University Press Archive.

Hazis, Faisal S. 2012. *Domination and Contestation: Muslim Bumiputera Politics in Sarawak*. Singapore: ISEAS.

Kua, Kia Soong. 2007. "Racial Conflict in Malaysia: Against the Official History". *Race & Class* 49 (3): 33–53.

Malaysiakini. 2002a. "Bintulu: The final nail in Snap's coffin? Pt II". *Malaysiakini Online*. www.malaysiakini.com.

———. 2002b. "Snap elected reps at the crossroads". *Malaysiakini Online*. www.malaysiakini.com.

————. 2003. "Tajem says PBDS must stay Dayak-based". *Malaysiakini Online*. www.malaysiakini.com.

————. 2011. "Cold shouldered, Snap quits Pakatan". *Malaysiakini Online*. www.malaysiakini.com.

————. 2014. "Adenan urged to clarify status of Teras, UPP". *Malaysiakini Online*. www.malaysiakini.com.

Milne, R. S., and K. J Ratnam. 1972. "The Sarawak Elections of 1970: An Analysis of the Vote". *Journal of Southeast Asian Studies* 3 (01): 111–122.

New Straits Times. 2011. 'The great Foochow factor'. 21 March.

Ngu, Ik Tien. 2011. "In Transition: The Political Culture of Chinese Society". unpublished paper.

Puyok, Arnold. 2005. "The 2004 Ba' Kelalan By-Election in Sarawak, East Malaysia: The Lun Bawang Factor and Whither Representative Democracy in Malaysia". *Contemporary Southeast Asia* 27 (1): 64–79.

Rabushka, Alvin, and Kenneth A. Shepsle. 2009. *Politics in Plural Societies*. New York: Pearson/Longman.

The Borneo Post. 2006a. "BN fields first ever Bisaya candidate in Batu Danau". 7 May.

————. 2006b. "Don't think twice, vote for BN: Kedayan chief". 12 May.

————. 2006c. "Constituency Factfile". 13 May.

————. 2006d. "Mawan confident no contest for all top posts in SPDP." 26 May.

————. 2006e. "PRS' 'censured' president says he is doing the right thing and he is always an optimist". 30 June.

————. 2006f. "We are all one community: Manyin". 10 September.

————. 2007a. "New young faces added in PBB line-up". 12 February.

————. 2007b. "Jimmy disappointed over Masing's statement". 20 April.

————. 2008a. "Chin's figures outdated, claim PBB Youth chiefs". 2 February.

————. 2008b. "Our sincere appeal". 7 March.

————. 2008c. "DAP proposes Iban deputy prime minister". 18 March.

————. 2008d. "Timik asked to thank SUPP". 27 March.

————. 2008e. "Thank you for your patience: Sagan". 29 March.

————. 2009. "Ok, Let's review Kidurong". 18 August.

————. 2010a. "Dr Rayong accepted into BN fold". 1 November.

————. 2010b. "Many countries want to copy BN unique system". 10 November.

————. 2011a. "Kidurong incumbent faces first timer from BN". 7 April.

————. 2011b. "CM wants Wong in cabinet". 22 April.

————. 2011c. "Sng tight-lipped on next move". 26 April.

————. 2011d. "PRS still can't believe they are now No. 2". 3 May.

————. 2011e. " 'No' to non-elected reps in cabinet". 4 May.

————. 2011f. "New ministry, seven new faces in Sarawak cabinet reshuffle". 29 September.

————. 2011g. "Dr Jerip: Dayaks not sidelined, but role must be recognized". 29 October.

————. 2011h. "Unseen hands at work?" 13 November.

————. 2011i. "Manyin applauds Riot's election as SUPP deputy chief". 13 December.

————. 2012. "Avoid racial issues in Hulu Rajang, says PRS president". 16 April.

————. 2013a. "Seven ministers from Sarawak". 16 May.

————. 2013b. "PBB prepares for future but no leadership change at TGA". 23 August.

————. 2013c. "Will UPP be SUPP without the 'S'?" 28 August.

The Malaysian Insider. 2011. "DAP picks Baru Bian for CM if Sarawak captured". *The Malaysian Insider Online*. www.themalaysianinsider.com.

The New Sarawak Tribune. 2010. "Partyless Sng keeps mum over his status". 20 November.

The Star. 2005. "SUPP gets ready to wrest back Kidurong from DAP". 8 May.

———. 2006. "Asst minister weds tycoon's daughter". 15 October.

———. 2011a. "10th Sarawak State Election/Nomination List 2011". 7 April.

———. 2011b. "10th Sarawak State Election 2011/Scorecard". 16 April.

———. 2011c. "Lawyer: Wong and team have jeopardised political standing in party". 10 December.

———. 2014. "Sarawak maintains cabinet line-up". 3 March.

The Sun. 2009. "Six new faces in Sarawak cabinet". 9 November.

The Sunday Post. 2010. "PRS insists on new rep for Pelagus". 28 November.

Utusan Malaysia. 2011. "New reality". 17 April.

Wicks, Peter. 1971. "The New Realism: Malaysia since 13 May, 1969". *The Australian Quarterly* 43 (4): 17–27.

4 West Kalimantan

Institutional and historical overview

4.1 Indonesia: the framework

The big question for Sarawak was whether the consociational polity can arrest fluid and multiple categories, shape them into a form suitable for power-sharing schemes and perpetuate one set of categories throughout each cycle of political competition. Sarawak defied the expectations, proving to be a polity in which each person identifies with at least two (sometimes three) ethnic categories and has incentives to switch between them on a frequent basis. The big question for Indonesia is in many ways the opposite: can the ethnic practice, fixed over time under the authoritarian regime, a) transform because of regime change, and b) remain fluid because of the institutional design?

The theory of ethnicity and the theory of coalitions (see Chapter 1) inform that multiple elections on multiple administrative levels should have the effect of multiple activated identities for each individual. Activation of a particular category depends, goes the assumption, on incentives embedded in each election (or non-electoral political event). So goes the hypothesis for this part of the analysis as well: this part of the work is committed to search for proof of multiplicity of identities activated due to differing incentives of different elections. West Kalimantan experienced not only a change of institutions, but also a remake of the constituencies, therefore establishing a set of potential brand new ethnic mosaics, distributions, power relations and minimum-winning coalitions. These elements offer a chance to activate multiple ethnic categories for each individual. However, Indonesia also put in place means which should discourage ethnic mobilization (local party ban, disapproval for explicit ethnic mobilization). Having this in mind, we should expect to observe difficulties in the activation of new categories.

The findings of this research go against what was hypothesized: a lot of ethnic inertia will be observed in West Kalimantan, suggesting that historically activated categories dominate politics and overshadow those categories that are mobilized within new administrative units and recently introduced direct elections. The inertia, and hence the relative fixedness of identities in West Kalimantan, will be attributed to both institutional factors (dual ticket in executive elections, informal ban on explicit ethnic mobilization), and to patronage (incumbent and family member advantage). Although new and alternative categories are activated in several regencies, in none of them does these new categories cut across the historically activated split between Malays (Muslims) and Dayaks (Christians). The change in subsequent gubernatorial elections and the activation of new, regional categories in the second gubernatorial election suggest that in the future more categories will be activated.

We will again begin with a historical recapitulation of category activation, but chiefly to show the stagnant set of categories activated continuously for an extended period of time. Historically three categories were activated: Malay, Dayak and Chinese (Davidson 2008, 31). The widespread understanding that these are the categories that matter was reflected in a study carried out by an Indonesian polling institute, Lingkaran Survei Indonesia (LSI). In May 2007 LSI conducted a study about influence of ethnicity on voters' behaviour in gubernatorial elections (Lingkaran Survei Indonesia 2008). The ethnic categories included in the study's design were: Malay, Dayak Chinese and others, as well as Muslims and Christians. The timing was carefully selected; the polling was conducted after candidates' nominations, but before elections. LSI found that for 56% of voters, the candidate's religion was an important factor in electoral choices; for 44% ethnicity [etnis] was an important factor (Lingkaran Survei Indonesia 2008, 6). The study also showed that although as many as 92% of voters were ready to accept a Muslim governor, only 65% would accept a Christian in this office. However, 72% of respondents were ready for a Malay governor, while 92% would have no problem with a Dayak governor. A Chinese governor, on the other hand, would be acceptable only to 49% (Lingkaran Survei Indonesia 2008, 7).[1]

LSI also asked about particular preferences for candidates in the then upcoming (2007) gubernatorial elections. The Christian and Dayak support was found to go to the only Dayak and Christian candidate, Cornelis. The Malay and Muslim vote was split between the other three candidates (all of them being Malays paired with Dayak Christians vice-governor candidates). Tellingly, among the Chinese and "Other", and "Other religion" (Hinduism, Confucianism, Buddhism) respondents, there were significant numbers of those who were undecided or did not wish to reveal their preferences.

The LSI poll was not sensitive to distinguish between Javanese, Madurese, Dayak and Malays among the Muslims; it did not distinguish between geographical divisions among Malays or Dayaks, or Protestants and Catholics, although in many regencies, as Tables 4.1 and 4.2 indicate, these categories have great potential for activation given their substantial numbers. Therefore, the research confirmed what was commonsensically known for decades: religion is a cleavage of West Kalimantanese politics. The poll was one of the many attempts devoted to proving that West Kalimantanese politics is driven by the phenomenon of Dayak-Malay-Chinese political competition – and it did. However, the study could not find anything that it was not looking for. Could it be that other categories were in play, but the researchers failed to capture them because of their research design?

My analysis of the 2007 gubernatorial election conducted in the next chapter is consistent with LSI's findings. In the 2007 gubernatorial election it was indeed the Malay-Dayak-Chinese division that explained candidates' mobilization strategies.

Table 4.1 Religious followers in West Kalimantan by regency/city according to Census Indonesia 2010

Regency/City	Islam	Protestantism	Catholicism	Buddhism + Confucianism*
Sambas	87.7%	1.8%	2.8%	7.6%
Bengkayang	36.4%	26.5%	30.8%	6.1%
Landak	16.0%	29.1%	54.3%	0.5%
Kayong Utara	95.5%	1.1%	0.5%	2.3%
Ketapang	69.6%	6.3%	21.5%	1.5%
Kubu Raya	82.7%	3.6%	5.5%	7.4%
Pontianak regency	80.2%	4.6%	6.2%	8.6%
Pontianak city	75.4%	5.0%	6.1%	13.3%
Singkawang city	53.8%	5.2%	8.0%	32.4%
Sintang	37.1%	23.5%	38.7%	0.6%
Melawi	51.9%	22.8%	24.6%	0.6%
Kapuas Hulu	59.5%	8.2%	32.0%	0.2%
Sekadau	38.3%	13.6%	47.1%	0.9%
Sanggau	33.1%	15.9%	49.2%	0.9%
WEST KALIMANTAN	**59.2%**	**11.4%**	**22.9%**	**6.1%**

Source: Badan Pusat Statistik (2010).

* Buddhism and Confucianism are professed only by persons of Chinese ancestry; the numbers of Confucianists are very low (less than 1% in the province) and as a separate category would be too small to study. Note, however, that many Chinese are Christians, and hence numbers in this column do not reflect the actual numbers of persons of Chinese descent.

Ecological inference of election results and religious followers' distribution confirms that the Muslim versus non-Muslim division was the cleavage of this election (see further chapters of this work). However, the point of this research is to find out whether the categories activated in the first gubernatorial election, which I argue is the baseline for ethnic politics in West Kalimantan, were again activated in the next election, and whether the same categories were activated in elections conducted at other administrative tiers. If the hypothesis that "more the elections, more the activated categories" is to be proved, it needs to be tested whether elections on other levels activate different categories. Therefore, legislative, presidential and *bupati* (regency head) and mayor elections will be analyzed. Political parties and ethnic organizations will be studied along the way to investigate their role in ethnic mobilization.

First, however, it will be shown how the categories LSI identified as important had earlier developed to be activated in West Kalimantan and be so prominent in the very first direct election of the governor in West Kalimantan. Recapitulation of path dependence analyses from secondary sources (mainly Davidson 2008; Tanasaldy 2012) will serve the purpose of showing how the categories were reinforced prior to the Reformasi era. In the next step, the analysis will move on to the legislative elections, mostly to show, however, that these, because of the extremely proportional electoral system in Indonesia, are hardly suitable for

Table 4.2 Ethnic (*suku bangsa*) categories in West Kalimantan by regency/city according to Census 2010[i]

Regency/City	Malay*	Java**	Madura	Dayak***	Chinese	Bugis
Sambas	82%	3%	0%	4%	8%	0%
Bengkayang	21%	10%	0%	57%	7%	0%
Landak	3%	3%	2%	88%	1%	0%
Kayong Utara	71%	13%	4%	1%	3%	4%
Ketapang	31%	10%	5%	46%	2%	0%
Kubu Raya	29%	17%	21%	6%	9%	11%
Pontianak regency	34%	6%	22%	13%	12%	9%
Pontianak city	34%	14%	12%	4%	19%	8%
Singkawang city	32%	9%	7%	8%	36%	0%
Sintang	16%	16%	0%	61%	2%	0%
Melawi	20%	5%	0%	68%	2%	0%
Kapuas Hulu	49%	3%	0%	43%	1%	0%
Sekadau	26%	9%	0%	58%	3%	0%
Sanggau	20%	9%	0%	62%	3%	0%
WEST KALIMANTAN	**34%**	**10%**	**6%**	**34%**	**8%**	**3%**

Source: Badan Pusat Statistik (2012).

  i The numbers do not come up to 100, as many respondents declared membership in other than categories included here. "Batak", "Daya" and "WNA" (*Warga Negara Asing*, or foreigners) in some *kabupaten* constituted up to 2% of the population. However, there were still dozens of other categories declared in the census.
  * For clarity, I grouped together three categories: "Malay", "Sambas Malays" and "Pontianak Malays". Sambas Malays are concentrated in the Sambas and Singkawang regencies of the province. Pontianak Malays are found mainly in Pontianak city and regency as well as Kubu Raya regency. In most other parts of the province, Malays identified themselves as Malays.
 ** This category does not include the Sundanese, i.e. people from the western part of the Java island. There are about 49,000 Sundanese people in West Kalimantan, distributed about evenly in all regencies.
*** This category is the sum of all categories that in the census had the prefix Dayak, except the "Dayak Melayu Pontianak" and "Dayak Melayu Sambas". See below for detailed information about Dayak composition in particular regencies.

drawing conclusions about the ethnic vote. However, party popularity can be well shown in the legislative election results, and this will be done against the background of religious composition of constituencies. Legislative elections also condition the executive candidates' nominations and hence, despite their limited conclusiveness, will be presented first.

Subsequently, the most promising executive elections will be discussed. Starting with the gubernatorial elections, I will demonstrate the reference point of this research – the assumed tripartite division. From there we will venture off to search for alternative activated categories in politics. Were there attempts to mobilize other categories and did they succeed? Has the ethnic split changed between the first (2007) and second (2012) gubernatorial elections? Which cleavages are to be found in regency (*kabupaten*) elections? Is the presidential election an ethnic matter and can it offer an opportunity to create new minimum-winning coalitions within the province? Given the vast area of the province and the decentralization that brought the *kabupaten* (regency) level to the fore, are we going to observe growing regionalism *within* West Kalimantan? How would it affect the historical religious or ethnic divisions? Hence, we expect to find the ethnic identity change by comparing:

1 Elections over time (e.g. the first direct election vs. the second direct election in case of executive positions).
2 Elections between administrative levels (e.g. the cleavage in the gubernatorial election and the *bupati* election within the same area).
3 First round and run-off, if it took place.
4 Regencies among themselves, to find patterns of candidates' profiles and minimum-winning coalitions.

4.2 Historical overview

Indonesia was conceived as a secular state, although the constitutional debates surrounding the work on the subsequent constitutional documents in Indonesia between 1945 and 1959 all had at their core the desire of the religious leaders to include Islam as the state religion (Indrayana 2008, 7). The discrepancy between the "nationalist" (i.e. secular) and the Islamic vision of the state was not a mere ideological dispute. Some leaders, most pronouncedly Hatta, saw concrete danger in pushing for an Islamic state. Christians, albeit a minority on average, are dominant groups in some of the provinces on the fringes of Indonesia and these provinces might oppose an Islamic state in the form of secessionist movements, went Hatta's argument. In the end, Indonesia was framed as a country of believers in one God – five religions were recognized and treated as equal, as stipulated in the national ideology called *Pancasila*. The language policy followed similar logic. The Malay language, at that point the *lingua franca* of the archipelago, was adjusted to accommodate local borrowings from the Dutch, and its standardized version was labelled as *Bahasa Indonesia*, or "the Indonesian language". The national language was promptly promoted as a token of unity and as a practical solution for facilitating nationwide education and media. The Javanese language was never seriously considered as alternative for the national language[2] (despite being the native language of more than 40% of the population, as compared to 3% in the case of Malay) in order to avoid the dissent of Indonesians in the outer islands ("the inner islands" being Java and Madura).

Despite the seemingly all-inclusive state ideology that was capable of accommodating all the ethnic categories (except for atheists and animists), the Chinese remained for decades on the fringes of the nation.[3] According to the 1945 constitution, there were *pribumi* (or native) Indonesians and "Indonesian citizens" who are not native; the latter category corresponding with the Chinese. Moreover, Sukarno gradually introduced economic limitations to Chinese activities, and schools using the Chinese medium were progressively closed. Dissimilationist measures against the Chinese were further maintained by Suharto; these included the necessity of possessing an additional identity document[4] as well as a distinct code on identity cards denoting persons of Chinese origin. At the same time, during the New Order, the Chinese were forced to change their names so that they resembled Indonesian names, and the Chinese language and script were eliminated from the public sphere. During Suharto's New Order the ethnic Chinese disappeared from politics. Indeed, the sole ethnic Chinese minister ever nominated only joined the very final Suharto cabinet, and served for a mere two months (Wibisono 2009).

The 1999–2002 constitutional changes also had as their focus the question of the position of Islam. However as before and because of the same concerns, Islam was not elevated to a position higher than other religions, but the issue accounted for heated discussions in the parliament during constitutional amendments debates (Mietzner 2008, 445). Note that it is the same 1945 constitution, albeit extensively amended, which is the valid document until now. The *Pancasila* still serves as the ideological basis for national harmony, as the moral common denominator between the religious groups and as the government-promoted, all-encompassing guideline for political parties and public personae.

In reference to ethnic relations the Reformasi, as the post–New Order period is known, had the most visible and pronounced impact on the position of the Chinese. Article 6 of the constitution, which had hitherto stipulated that the president must be "a native Indonesian citizen", was amended in 2001 and now reads: "Candidates for the President and Vice-President must have been Indonesian citizens since birth, must never have taken other citizenship of their own accord" (Indrayana 2008, 422). Furthermore, during the first three post-Reformasi presidential administrations (B.J. Habibie's, Abdurrahman Wahid's and Megawati Sukarnoputri's), the majority of the country's anti-Chinese regulations were abolished. Chinese press and other Mandarin-medium media are readily available and many public schools offer Mandarin lessons. Since 2006, Confucianism has been listed among the state's officially recognized religions (i.e. it is included in *Pancasila*). The Chinese New Year was celebrated as a public holiday for the first time in 2003. Each year the celebrations have grown larger and *Imlek* (as the festival is known in Indonesian) is now an important date in Indonesia's calendar of celebrations. Politicians participate in public events related to the occasion, seizing it as an opportunity to gain popularity among the Chinese voters.

This short account of the constitutional positioning of ethnicity and religion in Indonesia helps explain why "Chinese" and "Muslims" as categories may and often are assigned particular values in the political discourses in Indonesia, on both the national and sub-national levels. Unlike in Malaysia, Islam is not officially recognized as the state religion, but the day-to-day practice places Islam at a *primus-inter-pares* position among other religions. Also, the Chinese, despite the changing laws and declared inclusiveness of state ideology, invariably hold a distinct status in politics that reaches beyond the religious and regional divisions, and renders them a much more conspicuous category than any other in the ethnic mosaic of Indonesia.

Between 1945 and 2012 the Indonesian regime has transformed several times, both in terms of political freedoms and in terms of institutional technicalities.

Until 1955 the role of Parliament was carried out by a non-elected body, in which several parties and factions were represented, including nationalist, religious (Islamic and Catholic), communist and socialist (Mietzner 2008, 434). Between 1949 and 1955, prior to the first election, Sukarno and Hatta held political power. Sukarno was elected to his office by a committee[5] of no more than 20 individuals. The 1955 elections[6] results showed that almost 80% of voters supported one of four parties (PNI 22.3% of the votes, Masyumi 20.9%, NU 18.4%, PKI 16.4%),[7] while all other parties received less than 3% of the votes, and fewer than 10 seats in the 257-seat parliament. With no clear winner of the election and the necessity of building coalition cabinets, the work of the parliament and the cabinet was seriously hampered.

From 1959, during a period called Guided Democracy, the decision-making process was in the hands of Sukarno, who relied on the Communist Party of Indonesia (Partai Komunis Indonesia or PKI) and, albeit less with time, the military. In 1959 regional parties were forbidden and the existing ones had to dissolve according to presidential decree no. 7/1959; this regulation required each political party to have branches in at least a quarter of the provinces. Most of members of the defunct parties joined *en bloc* existing nationwide parties. The society under Guided Democracy was to undergo NASAKOM-isation, with NAS- standing for nationalism, A (from Indonesian *agama* meaning "religion") – for religious devotion and KOM- for communism (Anderson 1983, 485). The failed coup d'état of 30 September 1965 paved the way for General Suharto to take over power. By framing the 1965 affair as the making of the Communist Party, Suharto's regime legitimized not only deregistration of the party (at the time arguably the most popular one in Indonesia), but also the mass killings of the party members, sympathizers and assumed sympathizers. Suharto became the acting president in 1966 and took over as president in 1967.

The New Order, as Suharto's rule came to be known, did involve elections, but there was no actual political competition. Three political parties were allowed to exist, of which only the Golongan Karya (or the Golkar, "functional groups") had any real chance of winning elections. Although strictly speaking Golkar was not a party, since 1969 members of this organization were not allowed to be members in political parties, although many of them had been. Given the choice, many chose the Golkar membership over the party's. During the first election of the New Order in 1971, 10 parties competed with Golkar winning with more than 60% of the votes and only one other party (NU) receiving more than 10% of the votes. From then on, political pluralism was curbed even more and party politics became almost insignificant. In 1973 Muslim parties were forced to merge into one organization, Partai Persatuan Pembangunan (United Development Party, or PPP), and non-Muslim parties were put together to create Partai Demokrasi Indonesia (Indonesian Democratic Party, or PDI). Elections were rigged to ensure Golkar's continuous power, and Golkar in turn served as a machine for mobilizing mass support for Suharto. Only Golkar was allowed to have structures at the level below regency; PPP and PDI were not allowed to organize at the level of district or village (Kimura 2012, 51). This setup of limited political competition and lack of ideological pluralism, according to Suharto, was intended to serve political stability and unity, which were also secured, in any case, by the strong position of the military, both in politics and in administration. Public discourse on inter-ethnic, religious and race issues (referred to as SARA issues) was forbidden beginning in the early 1970s and *Pancasila* was deployed with increased force to fill the void of discussion. SARA issues remain sensitive matters in public opinion and on the occasion of each election candidates and parties are reminded to refrain from mobilizing along religious and ethnic lines (*Tribun Pontianak* 2010).

Golkar, argues Mietzner, although it "presented itself as culturally inspired by Islam, with many Golkar politicians in west Java, south Sulawesi or Sumatra promoting explicitly pro-Muslim policies" (2008,445), in Christian-majority areas cooperated closely with churches and missions and supported Christian leaders. Similarly, "Golkar leadership on Bali consisted almost exclusively of Hindus, reflecting the religious composition of the island" (Mietzner 2008, 445). Importantly, Golkar's strategy corresponded to the idea of a locally ethnic party. Golkar's past practice implies that it is a viable tactic of parties to on one hand mobilize local categories which otherwise have no particular party representation, but on the other hand to brandish a single, non-ethnic image nationally.

The end of the Suharto era came with the 1997–1998 financial crisis in Southeast Asia. In May 1998, the country's dire economic situation and the unstable political climate turned inner-city mobs against the Chinese middle class. Rapes, killings and arsons of Chinese estates drove dozens of thousands of Tionghoa (as the Chinese are referred to in Indonesian) out of the country for fear of their lives. Many Chinese had come to believe that they were accepted by the *pribumi* as compatriots because they spoke the national language, used Indonesian-sounding names and also made attempts to integrate into Indonesian society. Nevertheless, their economic status – a result of Suharto's policies – made them textbook scapegoats. Whether Suharto would have deemed standing up for the Chinese suitable or not, he had lost the authority necessary to order the protection of the minority by the armed forces,

The president was under pressure to step down, and did so. The interim president Habibie's administration paved the way for the first free elections held in Indonesia since 1955. A swift-running process of democratizing Indonesian political life began and, over the subsequent years, resulted in far-reaching constitutional changes, as well as the decentralization of power and renewed ideological pluralism.

4.3 Constitutional changes and institutional design in the post-Suharto era

Institutional changes after Suharto's downfall happened in stages. In the first step, party politics were liberalized and the first legislative elections took place in 1999. Until 2004, the president was elected in the national legislative assembly (Dewan Perwakilan Rakyat, or DPR), governors were nominated by the president and *bupatis* by the governor.[8] In the meantime party regulations changed thrice with the aim, on one hand, to eliminate regional parties, and on the other, to bolster strong nationwide organizations. The changing regulations resulted in parties being deregistered, re-registered under new names, splitting and, simply, dying out. The institutional outcome of these changes will be presented in the next paragraphs.

Elections are carried out every five years to four assemblies: Dewan Perwakilan Rakyat (People's Representative Council, or DPR), Dewan Perwakilan Daerah (Regional Representatives Council, or DPD), both on the national level; Dewan Pewakilan Rakyat Daerah (Regional People's Representative Council, or DPRD) on two levels: province and regency (*kabupaten*) or municipality. Polling to all assemblies takes place at the same time. Non-partisan candidates are not allowed to contest. Party lists are "open", i.e. voters choose particular candidates, and not merely a party. There is an electoral threshold of 3.5% to the DPR. Candidates for DPD are non-partisan individuals (not parties).[9]

Presidential candidates (president and vice-president, running on one ticket) must be nominated by political parties; eligible for nomination of a candidate pair are parties or coalitions of parties that either have 20% of the seats in DPR, or have received 25% of the popular vote in the previous election. A winning candidate team needs to obtain at least 50% of all votes and at least 20% of the votes in at least half of all provinces in Indonesia. If no candidate pair fulfils these conditions, a run-off takes place.

Governors and regents (*bupatis*) are elected according to similar rules. The first law introducing direct executive elections was passed in 2004, followed by an important amendment in 2008. Among others, the 2008 law introduced the possibility of independent (i.e. non-partisan) candidates in regional elections. Moreover, according to the 2004 bill, only 25% of the votes were required to win in the first round; this requirement was raised to 30% in the amended 2008 bill. The head of a regency (*kabupaten*) is called *bupati* and is elected along with a vice-*bupati* on one ticket; the same rule applies to governors in provinces. For both positions, candidates are nominated by parties or coalitions of parties that obtained at least 15% of the seats or 15% of the popular vote in the last legislative election on the respective administrative level. Non-partisan (individual) candidates may contest governor and *bupati* elections upon proving support for the candidacy of between 3% and 6.5% (depending on the population of the respective administrative unit, compare Law 12 of 2008) of the population in the province or regency. The winning pair of candidates must obtain at least 30% of the valid votes. If more than one pair obtains more than 30%, the one with more votes and wider distribution of votes wins.[10] If no pair reaches the 30% mark, a run-off takes place.

Direct executive elections became a symbol of commitment to democracy and observance of political fair play in Indonesia when the outgoing 2009–2014 DPR chamber voted in September 2014 to return to the system of local executives being elected by the respective legislatures. The background to promote such a bill was related to the concluding presidential elections. Six parties (Gerindra, Golkar, PPP, PKS, PAN and Demokrat) supported Prabowo Subianto in the election, who lost to Joko Widodo; the parties supporting Prabowo, however, constituted the majority in the newly elected DPR. Although the bill was overturned by a special presidential decree issued by the outgoing president Yudhoyono and supported by the 2014–2019 DPR chamber in January 2015 (*The Jakarta Post* 2015), which reinstated the elections, the fragility of Indonesia's democracy was demonstrated (Aspinall and Mietzner 2014). At the same time, the pressure on MPs to finally reinstate the direct elections clearly came from the bottom. Popular outrage at what voters perceived as depriving them of choice and influence over local and regional matters was so strong that any party could ill afford to find itself on the opposite site of this barricade (*The Jakarta Post* 2014). According to the newest regulation on regional and local elections, the elections are to be held simultaneously on the same date for all governors, mayors and regents across the country.[11]

As during the Old and the New Orders, Indonesian regulations allow for religious parties and hence, Muslim and Christian parties exist. Currently, parties are expected to follow one of two ideological tracks. The *Pancasila* track, also referred to as "nasionalis", is to be understood as "of the whole nation". *Pancasila* parties identify with no particular religion, but often underscore that they are "religious" (e.g. the Demokrat party, according to its official slogan brandished on its Web site and flyers, is a "nationalist-religious" party). Religious parties are predominantly those related to Islam, with a few small Christian parties. In the light of the theory followed in this work, we acknowledge all religious parties as ethnic parties. Ethnic parties that would wish to mobilize ethnic categories located in a specific area (e.g. Javanese, Bugis, Banjars, Minangkabau) are forbidden not because of an explicit ban, but because parties need to show a nationwide presence in order to be eligible for registration and participation in elections. As of 2011, party must have offices in at least three-fourths of provinces; have offices in at least 50% of *kabupaten* (regencies) of these provinces; have at least 1,000 members, or at least 1 member per 1,000 citizens in the respective areas – this regulation effectively precludes the existence of regional parties (except Aceh). In fact, all consecutive laws pertaining to political parties (1999, 2002, 2008, 2011[12]) not only upheld the regional party ban, but also subsequent amendments of the legislation raised the requirement of regional width of support. While according to the 1999 and 2002 laws, parties in order to be registered needed to demonstrate support in half of Indonesia's provinces, and within those provinces in half of the regencies, by 2008 it became 60% of provinces and 50% of regencies and 25% of sub-districts in those provinces. In 2011 the threshold was further raised to 75% of provinces and 50% of regencies.

Hence, there are no obstacles to creating ethnic parties like Partai Reformasi Tionghoa Indonesia (Indonesian Reformist Chinese Party), Partai Buddhis Demokrat Indonesia (Indonesian Buddhist Democratic Party), Partai Katolik Indonesia (Indonesian Catholic Party), Partai Kristen Nasional Indonesia (Indonesian Protestant National Party) (Ufen 2006, 10), as they wish to represent ethnic categories spread across the country (none of these parties was, however, successful) and in theory they can fulfil the legal requirements. However, more importantly in Indonesian conditions, parties wishing to mobilize based on ethnic categories concentrated in one area (e.g. an island, like the Madurese, part of an island, like the Javanese, or a local minority category, like the Dayaks) are forbidden. Therefore, unlike in Malaysia where parties were labelled with their ethnic appeal, in Indonesia parties will be tested based on their local popularity to establish whether they are ethnic, and one has to expect that the parties would deny such mobilization (if it were to exist). The definition deployed here of an ethnic party allows a party that is non-ethnic on the national level, but proves ethnic at the provincial level. These instances will be of utmost importance for this research.

In 2000 the constitutional provisions for regions [*daerah*] and their competences were amended to make provinces and regencies stronger in the spirit of regional autonomy [*otonomi daerah*] (Butt and Lindsey 2012, chap. 6). Already in 1999 a new Regional Government Law was passed which gave regencies and municipalities wide competencies. Provinces, according to the bill, "were not 'naturally' superior to counties [regencies] and cities in the new scheme of governance created by the legislation. They therefore could not trump the decisions or laws of local governments in the countries and cities within the province" (Butt and Lindsey 2012, 171). The later Regional Government Law, passed in 2004, reversed many of the 1999 provisions. Provincial heads were made responsible to the president, and governors also were given power to "guide and supervise governance in counties [regencies] and municipalities" (Butt and Lindsey 2012, 171). This law not only empowered provincial administration, but also gave the central government some control over sub-provincial governments.

The decentralization resulted not only in a reformed structure of power with competences moved down to provincial and regency levels, but also provided a strong incentive for the further creation of new provinces and regencies. While in 1999 in West Kalimantan there were six regencies (Sambas, Pontianak, Kapuas Hulu, Ketapang, Sintang, Sanggau), and one city: Pontianak, by 2009 there were 14 administrative units: Bengkayang regency separated from Sambas in 1999; Landak separated from Pontianak regency in the same year; Singkawang city split from Bengkayang regency in 2001, Melawi regency separated from Sintang in 2003 and Sekadau parted from Sanggau at the same time. Kayong Utara became a new regency after parting with Ketapang in 2007 and Kubu Raya was the last regency created (from *kabupaten* Pontianak, 2007). Creation of several more new administrative units is under way in West Kalimantan.

As Tanasaldy pointed out, in the case of Bengkayang, Landak and Singkawang, the new *kabupaten* were created to meet the demands of ethnic categories that wished for self-government (2012, 277–287). The ethnic rationale was not behind all regency formations (an example is Kubu Raya, which shares similar ethnic composition with its mother-regency Pontianak), however, further studies of this factor in the newly proposed *kabupaten* should reveal some interesting facts. The case of Sintang described in further paragraphs shows that ethnic considerations remain one of the important factors in deciding the shapes of new *kabupaten* and the formation of regencies may and arguably does represent a bargaining chip in inter-ethnic power negotiations. The proposal of formation of a new regency must be supported by the mother unit and its *bupati* (or governor, in the case of new provinces); naturally, the *bupati*'s consent for the proposal can be traded for e.g. electoral support for him in an upcoming election.

The current window of possibility for creation of new administrative units may not last; the question of how many provinces and regencies are needed in Indonesia and what purpose it serves to have more of them if their performance is not better than of that of the original bigger units is already being asked by both the government and non-governmental observers (*The Jakarta Post* 2012). Therefore, ethno-political entrepreneurs are likely to be currently working overtime to seize the present opportunity and achieve the goal of creating such a new administrative unit that will secure the highest possible returns for elites of an ethnic category that can be activated within a reasonably shaped regency. The next two decades in Indonesia should provide enough material to properly study the question of regency (province) formation as an incentive to mobilize new ethnic categories.

The traps of decentralization have been of concern to some political scientists, among which Vedi R. Hadiz is the most vocal on this topic. He notices that "local elites (especially at the sub provincial level [regency]) are intent on taking direct economic control, typically citing the injustice of past practices that allowed Jakarta to exploit Indonesia's vast riches. In the meantime, provincial authorities are stuck in the middle, struggling to retain some power and not to fall into the oblivion of political and administrative redundancy" (2004a, 705). According to Hadiz, Indonesian decentralization failed to deliver the benefits expected of it. Instead of curtailing predatory powers from Jakarta, decentralization resulted, among others, in "The emergence of decentralized, overlapping, and diffuse patronage networks built on the basis of competition for access and control over national and local institutions and resources; The rise of political fixers, entrepreneurs, and enforcers previously entrenched at the lower layers of the New Order's system of patronage" (2004b, 619).

Significant parts of the current research, albeit indirectly, are devoted to unveiling these elements of the networks that are based on ethnic relations and serve the purpose of ethnic mobilization. Equally importantly, in Chapter 5 I will trace some of these networks in West Kalimantan to the New Order's structures, in order to check if and how they were carried into the Reformasi era of freer political competition.

4.4 Ethnicity and inter-ethnic relations prior to Reformasi

During Dutch colonial rule most of West Kalimantan was ruled not directly by the Dutch, but by their proxies – in most cases the sultans. A few territories, among them the current area of Sintang, Melawi and Kapuas Hulu (Davidson 2008, 31–32), were directly ruled by the Dutch as there was no sultanate in the area on which the colonial government could rely. Tanasaldy indicates that not only did the areas of direct rule become a destination for Dayak migrants who wished to escape the oppressive control of the sultanate, but also that Dayaks governed directly by the Dutch advanced faster and lived more modern lives than those from indirect-rule regions (2012, 66). As Tanasaldy found, "L.H. Kadir calculates that the descendants of Kantuk Dayaks who decided to resettle into direct-rule areas in the beginning of nineteenth century are generally more successful than descendants of those who chose to stay under the sultanate's" (2012, 66). The Dutch rule marks the beginning of the "Dayak" identity, which imposed an idea of commonality on otherwise dispersed people of little unity. The Dutch first recognized the traditional law of *adat* and with it the "*adat* community" (Davidson 2008, 35–36).[13] Thanks to *adat*, Dayaks were assigned the status of a community with uniform traditions and law.

Malays' position as administrators and power bearers accounts for the main source of the original dichotomy between them and the Dayaks. However, at the time it was relatively easy for Dayaks to become Malays by converting to Islam. The conversion had both practical benefits (exemption from taxes, eligibility for positions) and less tangible consequences: a convert would cease to belong to a category routinely associated with backwardness and headhunting. In fact, according to Davidson already in the 17th century, converts to Islam considered joining the Muslim religious network as "progressive and modern" (2008, 24). The scope and geographical distribution of the conversions are less known, although it is likely that many of the current Melawi and Kapuas Hulu Malays are Dayak converts, as these areas are further in the interior than Sintang or Sanggau with a lesser Malay population (Tanasaldy 2012, 57). The argument goes: had the coastal Malays gradually migrated eastwards into the interior, their numbers would gradually decrease with distance from the coast. As this is not the case, it should be assumed that most of the interior Malays are Islamicized Dayaks. As was mentioned in Chapter 1, currently only relatively small numbers of the population identify as both "Dayaks" and "Muslims".

The Malay dominance was partly ended with the Japanese occupation during WWII. "1943 to 1944, the Japanese occupation forces kidnapped and summarily executed thousands of prominent local figures. Among those victims were all the sultans of West Kalimantan, who in many cases were killed together with their heirs and close relatives. Many Malay aristocrats, political activists, community leaders [were killed]" (Tanasaldy 2012, 74). As Tanasaldy (2012) and Davidson (2008) argue, the loss of leaders resulted in the later political weakness of Malays as a category in West Kalimantan.

The first Dayak organization called Daya in Action was established in 1945 in Kapuas Hulu with F.C. Palaunsoeka as its leader. A year later the organization was renamed as Partai Persatuan Daya (Dayak Unity Party, PD) (Davidson 2008, 37–38). While being led by Dayaks and for Dayaks, the party also attracted some Chinese who contested and won elections on the party's ticket, and PD maintained ties with Chinese organizations (Tanasaldy 2012, 113–119). In the 1955 general election the party obtained 146,054 votes, or 31% of the votes cast in West Kalimantan, and 169,222 votes to the provincial assembly in 1958 (Feith 1957, 65). PD's result was second best after Masyumi (modernist Islamic party). Presumably because of the PD's strength, the religious Christian party fared quite poorly in the elections in West Kalimantan, despite the province's significant proportion of Christians (only 2,500 votes in the 1955 election (Feith 1957, 69)). PD had 9 elected representatives in the national parliament, while Masyumi had 10. In the provincial assembly elected in 1958 PD was the biggest party with 12 seats (this time Masyumi won only 9). PD's leaders were *bupati* of five regencies in the province: Sanggau, Sintang, Kapuas Hulu and Pontianak (Davidson 2008, 42).

There was no "Malay" counterpart to the Dayak ethnic organization. Davidson attributes it partly to the localized networks of power that Muslim aristocrats exercised. Mempawah, Sambas and Pontianak were all of different ethnic lineage (Davidson 2008, 43) and, according to Davidson, an attempt to mobilize these regionally distinct Malays under the common umbrella would be futile. Moreover, Davidson argues that "exogenous national and religious" (2008, 44) forces influencing the local Malays were the reasons for weak politicization of Malayness, in contrast to the Dayak identity, which was independent from nationwide influences.

The Communist Party, despite being one of the pillars of the Sukarno regime, was not popular in West Kalimantan. Although the province had substantial numbers of Chinese, the ones who identified with the Communist Party were those who recently immigrated and were not Indonesian citizens and were more oriented towards Mao's China than Indonesian political affairs (Davidson 2008, 57). PKI obtained only 1.7% of the total vote in 1957 in West Kalimantan. PD was led by two leaders, F.C. Palaunsoeka and J.C. Oevang Oeray. The animosity between the two leaders led to the split of the party forces. After the presidential decree of 1959 that made regional parties illegal, Palaunsoeka joined the Catholic Party, which during the New Order was incorporated into the PDI. In 1987, Palaunsoeka was elected as a member of the DPR from PDI representing West Kalimantan province. Oevang Oeray and his faction of the PD in 1959 joined the Partai Indonesia (Partindo[14]). Oevang later served as governor in West Kalimantan and from 1970 until his death in 1986 was a member of DPR's Golkar faction (Magenda 2010, 81). Oevang Oeray had been able to win many Dayaks for Golkar, and after his passing the regime was seeking other ways to maintain its popularity among the Dayaks. In order to "institutionalize the delivery of Dayak votes" (Davidson 2008, 108), the regime established the Kenayatn Customary Council (Dewan Adat Kenayatn) in the Pontianak regency in 1985. The name referred to the dominant Dayak sub-category in the Pontianak regency.[15] Later similar institutions were established elsewhere and at lower levels of administrations, and as Dewan Adat Dayak became a powerful cultural and political organization.

In 1965 local military officers orchestrated a merger of the West Kalimantan Partindo branch – at the time the strongest party in the province – with IPKI (Ikatan Pendukung Kemerdekaan Indonesia, or Association of Indonesian Independence Supporters). The local IPKI was a party of limited popularity and overwhelmingly Malay leadership. As the numerous and influential Dayaks would overshadow the Malays' position in the party, the Malay elite took steps to reverse the merger by deciding to expel the former Partindo party members. The central leadership annulled the decision and installed a new local party leader who was to form the new party leadership. In the new line-up there were "28 percent Dayaks (6 out of 21). Most former senior Partindo members were excluded from the new structure, very likely in order to reinstate Malay supremacy in the party and to prevent the development of a stronger Dayak influence in the party leadership"

(Tanasaldy 2012, 131). In subsequent years, as Golkar was interested in expanding its base among the Dayaks and with many Dayak leaders disillusioned by their situation in IPKI, Golkar succeeded slowly to attract them from 1971 onwards (Tanasaldy 2007, 353).

However, significant pockets of Dayaks remained within the Catholic Party (PK) and IPKI. In 1973 these two parties (and three others, as mentioned earlier) merged into PDI. In West Kalimantan, despite those Dayak leaders who joined Golkar, it was the PDI that maintained a Dayak connotation. The "Dayak factor" helped the party during the 1997 elections in the province. After the regime removed Megawati from the PDI leadership in 1996,[16] PDI votes at the national level fell sharply from 14.9% in the previous election to 3.1% in the 1997 election. PDI votes in West Kalimantan that year, however, experienced a relatively small drop from 21.6% to a respectable 15.1%, the best showing among all provinces (Kristiadi, Legowo and Budi Harjanto 1997, 168).

To sum up, in West Kalimantan this first period after independence was marked by several features. Firstly, the Dayak category promptly and effectively organized politically. The category had been constructed during the Dutch rule around two elements: subordination to the Malays and the uniform *adat* law. With the political party and strong leadership, the category gained prominence exceeding that of the Malays. Secondly, Dayaks cooperated well with local Chinese also on the political scene; the Chinese, most of whom were not citizens, did not engage in nationwide Chinese political institutions. Furthermore, the party regulations forced the Persatuan Daya and its leaders to look for national party banners that would fulfil the legal requirements of nationwide parties, but in the province could carry the Dayak interests. The idea of a locally Dayak party as a *modus operandi* was born; however, personal and ideological differences between leaders of the PD led to split in the elites. Finally, there was no Malay party – whether as an explicit organization prior to 1959 (when it would have been legal), or as a local branch of a nationwide party after 1959.

Throughout the New Order period, frequent violent episodes between Dayaks and Chinese, as well as Dayaks and Madurese, took place. They will be recounted here briefly to point out their importance in galvanizing some West Kalimantanese categories in the period when institutionalized politics were devoid of ethnicity. The bloody clashes of 1999, 2000 and 2001 between Dayaks and Madurese as well as Malays and Madurese, although already taking place during Reformasi, will also be dealt with here. The 1965–1967 military operations in West Kalimantan were devised to eliminate the communist/Chinese/Sarawak PGRS guerilla movement (Pasukan Gelilya Rakyat Sarawak, or Sarawak People's Guerrilla Force). For this purpose, "the military sought to provoke Dayaks to attack ethnic Chinese, and thereby drive them from the interior to the coast where they could be controlled, counted and prevented from providing supplies to the rebels" (Davidson 2008, 65). Several acts of massacre between Chinese and Dayaks took place in 1967 in areas of the current Bengkayang and Pontianak districts (Davidson 2008, 67). These events, although clearly a link in the chain of anti-communist violence launched by Suharto, were portrayed in official accounts as a primordial ethnic conflict and "suggested that there was a deep-rooted and uniform animosity toward local Chinese. In doing so, [they] ignored the immense linguistic and cultural variations among Dayaks" (Davidson 2008, 68).

Although the army was the main instigator of these events, Dayak leader and former governor of West Kalimantan Oevang Oeray was also involved in the Chinese–Dayak clashes. Oeray calculated that driving the small-scale Chinese traders out of the Dayak-dominated area would open the field of economic activities to the Dayaks (Davidson 2008, 69). Despite having lost the governorship, Oeray was still a renowned figure and local leaders from the Bengkayang area where the killings were happening approached him in Pontianak. Soon after the meeting "a 'declaration of war' against the Chinese was announced" (Davidson 2008, 69). Dayak leaders expressed their support towards the armed forces and called on their fellow Dayaks to assist the military in its actions against the guerrillas. The Suharto regime marketed the tragic events in West Kalimantan between 1967 and 1972 as a rebellion with two villains. By "labeling all Chinese as rebels or potential rebels and all Dayaks as primitive headhunters" (Davidson 2008, 76), the regime presented the two categories as prone to hostility and violence against each other.

Davidson concludes that Oeray's followers among the Dayak ended up as participators in the anti-Chinese actions and were "ethnic extremists", while the Catholic Dayaks who sided with Palaunsoeka refrained from violence (2008, 72). Foremost, shows Davidson, although the Dayaks did not take over businesses from the fleeing Chinese, their ambition to expand into the trading field emerged. When the Madurese started to engage in trade activities previously operated by the Chinese, the Dayak-inspired violence against the Madurese broke out just weeks after the anti-Chinese massacres (Davidson 2008, 73). Davidson found evidence of Dayak–Madurese clashes taking place in different locations at least five times between 1969 and 1983 (2008, 89–90). The deadliest of the Dayak–Madurese clashes, however, elsewhere referred to as "communal war" (Peluso and Harwell 2001, 84) took place in 1997 in Pontianak regency, claiming the lives of 400 to 700 people, mostly Madurese (Davidson 2008, 102).

The 1999 Malay–Madurese clashes in Sambas and the subsequent exodus of all Madurese from the region took place after Suharto's fall. As Davidson noted, they were Malays' response to the Dayaks' newly acquired prominence on the political scene in the first years of Reformasi, during which Dayaks acted as the only rightful indigenous category worthy of political power (2008, 126). Malays' claim to indigeneity in the province was emphasized in their attack on the Madurese in Sambas. Another incident of Malay–Madurese riots linked closely to an intense political situation[17] took place in Pontianak municipality in October 2000 (Davidson 2008, 156–160). This way, Dayaks, Malays and Chinese became categories that in political practice came to be associated with entitlement to power in the province.

Violent episodes in West Kalimantan never ran along religious lines; none of the parties in the multiple clashes was framed as "Christians", "Muslims" or "Buddhists". Each time the clashes were "Dayak"–"Chinese", "Dayak"–"Madurese" or "Malay"–"Madurese" clashes. Therefore, we need to acknowledge two important elements of the ethnic puzzle. Firstly, the indigenous versus newcomer cleavage came out strongly reinforced from these clashes. Paradoxically, however, against the background of the more recent incomers the Madurese, the Chinese are now considered *putra daerah* (lit. sons of the region, or natives), and, as it will be shown later, they are entitled to share power in West Kalimantan, signifying their rightful residence in the province. Secondly, these conflicts led to the activation of categories characterized by no straightforward equation sign between "Christians" and "Dayaks", or "Muslims" and "Malays". Followers of Islam comprise multiple categories (including Dayaks!), and Malays are ever decreasing in numbers against Muslims of other backgrounds.

Next to the violent episodes, ethnic organizations helped galvanize some of the activated categories in West Kalimantan. The *adat* law was shown to serve as an ethnic category membership rule since the times of the Dutch. Starting from the mid-1980s, *adat*-based ethnic organizations were created, which reinforced the "adat communities" and serve as mobilization tool for the categories they claim to represent. The Majelis Adat Dayak (MAD, Dayak Customary Council, with its regency and district chapters) was established in 1985. The Majelis Adat Budaya Melayu (Malay Customary and Cultural Council, MABM) was launched in 1997 (Tanasaldy 2012, 303), while the Chinese followed suit in 2005 with the Majelis Adat Budaya Tionghoa (Chinese Customary and Cultural Council, or MABT) (Hui 2011, 292).[18] These organizations participated and represented their respective categories in reconciliatory events following violence; they are present on occasion of festivals and their leaders speak for their communities in matters of political importance. Most importantly, the *adat* councils are significant para-legal institutions as they exercise the customary law and deliver sentences in cases of breaching the customary rules of their respective communities. On a day-to-day basis, the said organizations also serve as leader-grooming machinery.

The position of the sultans (or *rajas*) is an identity boost for the Malays. Their role as the embodiment of Malayness was underscored during a controversy sparked by the Chinese mayor of Singkawang, Hasan Karman. In 2010 a 2008 academic text by Karman was found, in which he described 17th-century Sambas and Sukadana Malays as, among others, robbers.[19] The issue caused riots in the city and in order to resolve the tension, an *adat* ceremony was arranged. Hasan Karman paid an official visit to the Sultan of Samabas (*Pontianak Post* 2010b). During the ceremony he read his apology, addressed to all the *rajas* of West Kalimantan and the entire Malay community.

Muslim Dayaks established their own organization, Ikatan Keluarga Dayak Islam (Association of Muslim Dayak Families, IKDI) in May 1999. IKDI, argues Tanasaldy, was "formed to show that one could be both Dayak and Muslim at the same time" (2012, 270). The first chairman of the West Kalimantan PDI-P (Partai Demokrasi Indonesia-Perjuangan, or Indonesian Democratic Party-Struggle), Rudy Alamsyahrum, was an IKDI member. The Madurese established a similar body, Ikatan Keluarga Besar Madura (Association of Madurese Great Families, IKBM). Not being equipped with the power of *adat*, these organizations cannot match the importance of the *adat* councils and arguably have a weaker potential to bolster identities attached to the communities they aspire to represent.

The *adat* councils and ethnic associations try to maintain an apolitical image, which is difficult at best, as many regents simultaneously hold positions as heads of these organizations. MAD secretary Yakobus Kumis explicitly denied MAD being a political organization (interview by the author, 2 June, 2011). However, MAD issues its recommendation for candidates in elections. Prior to the 2012 gubernatorial election, Cornelis, likely concerned that MAD's endorsement may limit his appeal among non-Dayaks, requested that local branches of the Council refrain from issuing written letters of support for him in electoral campaign (Kompas 2012). Cornelis argued that gubernatorial election is not an election for the leader of an ethnic group (in this case Dayak), but the leader of the province, and in fact, a representative of the government. The criticism of MAD's politicization also came from Milton Crosby, himself the head of the Dewan Adat Dayak in Sintang and regent of Sintang; Milton argued that MAD should refrain from politics because it is not a political organization, but a cultural one (*Borneo Tribune* 2012). The IKBM provincial chief in an interview stated that while there was no written declaration, his personal attendance and speeches at Cornelis' rallies were tantamount to official endorsement (Sarumli Seneh, interview by the author, 13 September, 2012).

Two further Dayak organizations, the Institute Dayakology and a micro-credit foundation Pancur Kasih must be mentioned. Pancur Kasih was established in 1981 by Dayak and Catholic intellectuals, most prominently AR Mecer. In 1990 a group of activists from Pancur Kasih ventured to start Institute Dayakology Research and Development (IDRD), a cultural organization aimed at preserving the knowledge and traditions of Dayaks. *Kalimantan Review* is a monthly published by IDRD that covers a wide variety of topics related to Dayak culture, society, politics and development. Many activists from Pancur Kasih and IDRD entered politics: AR Mecer became a member of the Majelis Permusyawaratan Rakyat (People's Consultative Assembly, or MPR[20]) as the Dayak minority representative from Kalimantan[21] for the 1999–2004 term, Maria Goreti was a first- and second-term DPD member, Erma Suryani Ranik was a second-term DPD member and the 2009–2014 MP from the Demokrat party.

These organizations are introduced here to accentuate that, next to violent episodes that tend to reify certain ethnic categories, institutionalization of ethnicity happens through these councils and associations. Political links are maintained between executive offices and heads of these organizations. Membership in an ethnic organization gives a single ethnic label to a candidate and underscores his membership in this particular ethnic category. In the context of implicit ethnic mobilization, membership in these organizations is a strong tool to convey a message of one's chosen category.

The first years after Suharto's fall should be seen as a new political reality within the old institutions. Although Indonesia was still a highly centralized state, new social forces were starting to exercise their pressure on the government. New political parties were weak, but between old political organizations the competition was fierce and Golkar ceased to enjoy some of its privileges (but by no means all of them). Local executive positions, albeit still filled through indirect elections and with involvement from Jakarta, now could be manipulated locally. Dayaks turned out to be extremely successful in lobbying for their candidates in *bupati* (regent) nomination processes. The election process for the Sanggau regent is exemplary. Candidate selection for the post had begun in early 1998, with the Sanggau DPRD collecting propositions for candidates. At some point the list included as many as 40 candidates, half of them being personae with military background, as was usual during the New Order (Tanasaldy 2012, 261). During the months of completing the candidates list, Suharto's regime collapsed. The governor, who used to have a say in the *bupati* selection, was now under pressure from newly emancipated masses to let the DPRD elect according to the legislators' preference. The short list presented to the minister of internal affairs in September 1998 included four names – Mickael Andjioe, Benedictus Ayub, Donatus Djaman and Setiman Sudin (all Dayaks except for Sudin, a Malay). As Tanasaldy concludes, the fact that the minister struck off one of the Dayak names from the list indicates that there was a scheme devised to let a Dayak candidate win: by eliminating one Dayak candidate, the split of Dayak votes was avoided (2012, 264). Mickael Andjioe was elected in the end. Had the election taken place a few months earlier, observes Tanasaldy, the governor's man would have become the *bupati*. In the new reality, his name was not even presented for approval to the minister.

Tanasaldy also gives a detailed and interesting account of ethnic bargaining on the occasion of the 1999 Pontianak regency *bupati* election (2012, 265):

> The election committee from the DPRD tried to keep an ethnic balance between the Dayaks and the Malays in the nominations throughout the selection process as a way of avoiding ethnic strife. The committee tried to downplay polarization between competing ethnic groups, for example, by using the terms coastal (*pantai*), inland (*pedalaman*), and migrant (*pendatang*) communities instead of more sensitive ethnic terms, although these terms were clearly referred to Malays, Dayaks, and migrants respectively. The regime also denied having considered the ethnic factors in the process of nomination, although the way they preserved the ethnic balance throughout the nomination process showed the contrary.
>
> (2012, 264)

The final list included three Malay candidates, three Dayaks, and the seventh one represented a "migrant community" (Tanasaldy 2012, 266). This 3–3–1 combination was a modification of the original set: four Malays, two Dayaks and one non-Kalimantanese Muslim. After reducing the 3–3–1 list to 2–2–1, the names were submitted to the interior minister (Tanasaldy 2012, 265). The observance of ethnic proportions served to satisfy the pressure from the ground. After one of the Dayak names on the list was removed by the minister,[22] a group of 300 Dayaks burned down the Pontianak regency DPRD building (Tanasaldy 2012, 266–267). Reconciliatory steps that followed this event led to the election of

Cornelius Kimha as the regent of Pontianak. Soon after, Landak with its Dayak majority was separated as a regency from the coastal Malay-majority part of the *kabupaten* Pontianak.

An analogous approach to candidate selection was taken when the provincial DPRD was deciding on West Kalimantan regional representatives to the People's Consultative Assembly (Majelis Permusywaratan Rakyat, or MPR) in 1999. Five seats were to be filled and the regulation stated that the five candidates who received the most votes in DPRD would automatically be elected as Utusan Daerah (Representatives of Regions) members. The assembly members agreed on a compromise solution that among the five representatives there would be two Malays, two Dayaks and one Chinese. The result of the actual voting brought, however, only one elected Christian Dayak, along with two Malays,[23] one Dayak convert to Islam (Zainuddin Isman) and one Chinese. Tanasaldy refers to this situation as a failure to elect two Dayak candidates, arguing that a Muslim Dayak who does not explicitly identify as a Dayak is in fact a Malay, or in any case not a Dayak (2012, 269). Despite attempts to change the outcome, no alternative solution was found and the 3–1–1 set of representatives was sworn in as MPR members. Dayak groups protested the result for months and the DPRD was under pressure to change the decision, but the voting was never repeated.

The 2002 last indirect gubernatorial election in West Kalimantan also bore characteristics of elite power sharing. Aspar Aswin, an army general and native of East Kalimantan, had been the governor for 10 years. The 2002 election was mainly a choice between Usman Ja'afar-LH Kadir (the first a Malay, the latter a Dayak) and a Golkar nominee Gusty Syamsumin with Sebastian Massardy Kaphat (also a Malay-Dayak pair). The two other pairs were Djawari with Rudy Alamsyahrum (Malay with Muslim Dayak) and Henri Usman with Michael Oendoen (Malay and Dayak). The fact that all these pairs included a Malay and a Dayak (as governor and vice-governor, respectively) was, argued Davidson, stipulated in a tacit agreement reached on the basis of an argument that since the last local governor (Oevang Oeray) had been a Dayak, the first post–New Order one should be a Malay paired with a Dayak deputy (2008, 160). Ja'afar Laurentius Hermanus (LH) Kadir won the election in the provincial DPRD.

Therefore, in the ethnic bargaining on the provincial level, three ethnic categories were seen as eligible to share power: Malays, Dayaks and Chinese. Malays and Dayaks, went the logic, should split the governor and vice-governor posts. Five MPR seats were to be split 2–2–1, the third category being the Chinese. These deals had nothing to do with the numerical strength of categories, or with their precise membership definition of the categories. Javanese were not included in the MPR regional candidate line-up, although according to the 2000 census the numbers of the Chinese and Javanese in West Kalimantan were about even.[24]

The previous paragraphs focused on ethnic dynamics in West Kalimantan prior to the introduction of the new institutional setting with directly elected executive heads. I showed that throughout this entire period since the Dutch time, the ethnic practice in West Kalimantan involved very few categories. Three categories:

"Malays", "Dayaks" and "Chinese" were activated consequently and in many arenas: through their ethnic organizations, political parties, violence. In the most recent violent episode a new divide appeared: Malay–Madurese. Freer political life between 1999 and 2004 combined with indirect election to executive positions introduced the idea of elite bargaining leading to power sharing. Dayaks, Malays and Chinese were, according to these political deals, eligible to share power in the province.

This apparent simplicity of ethnic practice in West Kalimantan calls for a revision. I will therefore look into the most recent elections and other political happenings to establish if and which other categories are activated. Most importantly, I will attempt to match the category activation to particular elements of the political setting in Indonesia.

Notes

1 52% of Malays were not ready for a Dayak governor, and 69.8% of Malays were not ready for a Chinese governor. 16% of Dayaks would not accept a Malay governor, and 39.5% would have a problem with a Chinese in this position (Lingkaran Survei Indonesia 2008, 8).
2 See Paauw (2009) for analysis of the language policy in Indonesia.
3 See Prasad (2013) for the chronological dynamics of Chinese politics in Indonesia.
4 The *Surat Bukti Kewarganegaraan Republik Indonesia* (Letter of Proof of Indonesian Citizenship, or SBKRI) was introduced in 1958 before the citizenship question was resolved by the treaty with China; by any logic, the document should have become redundant and obsolete upon formal acquisition of citizenship and possession of an identity card, but remained in use throughout the New Order, after most of the Chinese obtained Indonesian citizenship.
5 Committee for the Preparation of Indonesian Independence, created by the Japanese in August 1945 (Anderson 1983, 480).
6 In September 1955 the voters elected the legislative assembly members; in another election in December that year, the Constituent Assembly was elected; in this election the support of each party was virtually the same as in September. In 1957 an election to provincial assemblies took place.
7 Clifford Geertz (1963) and later others used the concept of *aliran* (Indonesian for "streams") to explain cleavages in the Indonesian society in 1950s. The main political parties of the time went along, argued Geertz, the corresponding groupings in the society. Ufen (2006, 2008) tracks back the current parties' and party system's relation to the *aliran*. The *aliran* system combines ethnic (e.g. religious) and non-ethnic categories (e.g. peasants).
8 Between 1957 and 1959 governors and *bupatis* were elected through the respective parliaments; from 1959 to 1999 these positions were filled by nomination. *Bupati* candidates were proposed by the regency legislative assemblies, sanctioned by the Home Affairs Minister and elected by the assembly (Tanasaldy 2012, 105).
9 Article 11 of Law 10 of 2008 on General Elections for Members of the DPR, DPD and DPRD.
10 Article 107 of Law 12 of 2008 on General Elections for Members of the DPR, DPD and DPRD.
11 Government Regulation in Lieu of Law No. 1 of 2014 on the Election of Governors, Regents and Mayors; article 201 specifies how the varying terms of current officeholders will be adjusted to accommodate the new regulation.
12 Law 2 of 1999 on Political Parties; Law 31 of 2002 on Political Parties; Law 2 of 2008 on Political Parties; Law 2 of 2011 amending the Law 2 of 2008 on Political Parties.

13 Lukito (2012) gives a good account of the historical context of the *adat* as well as its current position in the Indonesian legal system.

14 Partindo was newly established (1958), secular, leftist and suitable for both the Chinese and Dayak supporters, whom Oeray wished to drag along from PD. One obvious reason for this decision was that merging with Partindo required the smallest political sacrifice; PD could preserve its structure and personnel because Partindo still had no presence in West Kalimantan. In fact, as far as West Kalimantan was concerned, the merger would be a mere change of party name, from "PD" to "Partindo" (Tanasaldy 2012, 111). Oevang Oeray regretted the leftist association later on and sided with the military during the 1967 raids against the communists (Davidson 2008, 69).

15 Pontianak regency at the time included the current Landak regency.

16 Between 1993 and 1996 PDI was becoming more and more critical of the government and chose Megawati Sukarnoputri as the chairperson. In 1996 under direct pressure from the regime, Megawati was removed from her position as chair of PDI (Ufen 2002, 346–350). After Suharto's fall, she established PDI-Perjuangan.

17 Davidson argues that the then-governor, Aspar Aswin, instigated the October 2000 Pontianak riots to derail his impeachment. In 2000 several factions of the provincial DPRD attempted to remove Aspar from his office (2008, 156–160). They failed in their first attempt because of procedural inaccuracies, but pressure from the elites and students' demonstrations continued. The riots in Pontianak effectively stopped the impeachment threat. Interestingly, in the 2004 legislative election, Aspar and his wife, Sri Kadarwati, were both elected members of DPD from West Kalimantan.

18 Prior to MABT, the Chinese were organized in associations, of which the most prominent was Bhakti Suci, led by Budiono Tan. Tellingly, it was Tan who was elected as Chinese representative of West Kalimantan to MPR in 1999.

19 According to *Pontianak Post* (2010a), the controversial excerpt was "In the 17th century, the Malays were involved in trade [*perdagangan*] and robbery [*perampokan*] and strengthened their position in the estuaries along the coast of West Kalimantan by setting up several centers and maritime bases. Among them, the kingdom [*Kerajaan*] of Sambas and Sukadana developed particularly well. From there, the network of trade, taxation and robbery expanded, especially to remote areas where the Malay leaders married Dayak daughters and strengthened their position in the areas of their in-laws".

20 The body composed of the DPR and, until 2004, 200 indirectly elected representatives of regions and social groups. Since 2004, MPR consists of representatives to the DPR and Dewan Perwakilan Daerah (DPD).

21 *Utusan Golongan* were seats assigned to particular social categories, including ethnic minorities. Mecer represented Dayaks from all Bornean provinces. This seat is not to be confused with the *Utusan Daerah*, seats in the MPR that were (five for each province) assigned to regions.

22 It was the later first directly elected West Kalimantan governor, Cornelis. According to information obtained by Tanasaldy, Cornelis was removed from the list as inexperienced and not senior enough (2012, 265).

23 One of whom was Oesman Sapta Oedang, elsewhere identified as of Bugis origin (Lingkaran Survei Indonesia 2008, 4).

24 There were 352,937 Chinese and 341,173 Javanese in the province in 2000, according to the census (Badan Pusat Statistik 2000). Significantly, by 2010 there were 427,221 Javanese, and 358,451 Chinese (Badan Pusat Statistik 2012).

References

Anderson, Benedict R. O'G. 1983. "Old State, New Society: Indonesia's New Order in Comparative Historical Perspective". *The Journal of Asian Studies* 42 (03): 477–496.

Aspinall, Edward, and Marcus Mietzner. 2014. "Indonesian Politics in 2014: Democracy's Close Call". *Bulletin of Indonesian Economic Studies* 50 (3): 347–369.

Badan Pusat Statistik. 2000. *Kalimantan Barat Dalam Angka 2000*. Pontianak: BPS.

———. 2010. *Penduduk Menurut Wilayah dan Agama yang Dianut. Provinsi Kalimantan Barat. Badan Pusat Statistik Online*. www.sp2010.bps.go.id.

———. 2012. *Suku Bangsa. Sensus Penduduk 2010*. N.p. Unpublished.

Borneo Tribune. 2012. "Kembalikan Fitrah DAD". 5 September.

Butt, Simon, and Tim Lindsey. 2012. *The Constitution of Indonesia: A Contextual Analysis*. Oxford and Portland, Oregon: Hart Publishing.

Davidson, Jamie S. 2008. *From Rebellion to Riots: Collective Violence on Indonesian Borneo*. New Perspectives in Southeast Asian Studies. Madison: University of Wisconsin Press.

Feith, Herbert. 1957. *The Indonesian Elections of 1955*. Modern Indonesia Project. Ithaca, NY: Cornell University Press.

Geertz, Clifford. 1963. *Peddlers and Princes; Social Change and Economic Modernization in Two Indonesian Towns*. Chicago, IL: University of Chicago Press.

Hadiz, Vedi R. 2004a. "Decentralization and Democracy in Indonesia: A Critique of Neo-Institutionalist Perspectives". *Development and Change* 35 (4): 697–718.

———. 2004b. "Indonesian Local Party Politics". *Critical Asian Studies* 36 (4): 615–636.

Hui, Yew-Foong. 2011. *Strangers at Home: History and Subjectivity among the Chinese Communities of West Kalimantan, Indonesia*. Leiden and Boston: BRILL.

Indrayana, Denny. 2008. *Indonesian Constitutional Reform, 1999–2002: An Evaluation of Constitution-Making in Transition*. Jakarta: Penerbit Kompas Buku.

Kimura, Ehito. 2012. *Political Change and Territoriality in Indonesia: Provincial Proliferation*. London/New York: Routledge.

Kompas. 2012. "Cornelis Minta Ketua Dewan Adat Dayak Tak Beri Dukungan Tertulis". 22 May.

Kristiadi, J., T.A. Legowo and N.T. Budi Harjanto. 1997. *Pemilihan Umum 1997: Perkiraan, Harapan, Dan Evaluasi*. Jakarta: Centre for Strategic and International Studies.

Lingkaran Survei Indonesia. 2008. "Faktor Etnis Dalam Pilkada". *Kajian Bulanan* 09. www.lsi.co.id.

Lukito, Ratno. 2012. *Legal Pluralism in Indonesia: Bridging the Unbridgeable*. London/New York: Routledge.

Magenda, Burhan Djabier. 2010. *East Kalimantan: The Decline of a Commercial Aristocracy*. Singapore: Equinox Publishing.

Mietzner, Marcus. 2008. "Comparing Indonesia's Party Systems of the 1950s and the Post-Suharto Era: From Centrifugal to Centripetal Inter-Party Competition". *Journal of Southeast Asian Studies* 39 (03): 431–453.

Paauw, Scott. 2009. "One Land, One Nation, One Language: An Analysis of Indonesia's National Language Policy". *University of Rochester Working Paper* 5 (1): 2–16.

Peluso, Nancy Lee, and Emily Harwell. 2001. "Territory, Custom and the Cultural Politics of Ethnic War in West Kalimantan, Indonesia". In *Violent Environment: Social Bonds and Racial Hubris*, edited by Nancy Lee Peluso and Michael J. Watts, 83–116. Ithaca, NY: Cornell University Press.

Pontianak Post. 2010a. "Maaf Diterima, Agendakan HK Temui Pangeran Ratu". 6 June.

———. 2010b. "Permohonan Maaf Resmi Dimatangkan". 3 June.

Prasad, Karolina. 2013. "Nationalising States and Nationalising Policies in Southeast Asia". In *New Nation-States and National Minorities*, edited by Julien Danero Iglesias, Nenad Stojanovic, and Sharon Weinblum, 125–150. Colchester: ECPR Press.

Tanasaldy, Taufiq. 2007. "Ethnic Identity Politics in West Kalimantan". In *Local Politics in Post-Suharto Indonesia: Renegotiating Boundaries*, edited by Henk Schulte Nordholt and Gerry van Klinken, 349–372. Leiden: KITLV.

———. 2012. *Regime Change and Ethnic Politics in Indonesia: Dayak Politics in West Kalimantan*. Leiden: KITLV.

The Jakarta Post. 2012. "How many provinces does Indonesia need?" 20 April.

———. 2014. "Supporting direct elections, Golkar faction changes party slogan". 8 December.

———. 2015. "Direct elections reinstated". 21 January.

Tribun Pontianak. 2010. "Baliho Perdamian Pilkada Tersebar". 20 April.

Ufen, Andreas. 2002. *Herrschaftsfiguration Und Demokratisierung in Indonesien (1965–2000)*. Mitteilungen Des Instituts Für Asienkunde, Hamburg. Hamburg: IFA.

———. 2006. *Political Parties in Post-Suharto Indonesia*. GIGA Working Papers. Hamburg: GIGA.

———. 2008. "From Aliran to Dealignment: Political Parties in Post-Suharto Indonesia". *South East Asia Research* 16 (1): 5–41.

Wibisono, Christianto. 2009. "Learning from Malaysia's mistakes". *Inside Indonesia* Online. www.insideindonesia.org.

5 Identity activation in the new institutional setting

5.1 Political parties

Prior to the first free election after Suharto's fall in 1998, the party scene in Indonesia underwent an enormous change; just two years earlier voters were choosing from three parties, now from 48. The local West Kalimantanese party mosaic reflected the province's specificity, and so did the 1999 election result. Nationwide, the winners were Partai Demokrasi Indonesia – Perjuangan (Democratic Party of Indonesia – Struggle or PDI-P), Golkar, Partai Kebangkitan Bangsa (National Awakening Party, or PKB), Partai Persatuan Pembangunan (United Development Party, or PPP), Partai Amanat Nasional (National Mandate Party, or PAN) and Partai Bulan Bintang (Crescent Star Party, or PBB) in this order. The first two are *Pancasila*-ideology, catch-all parties; the latter four are Islamic ones.[1] In West Kalimantan, however, the top six parties were Golkar, PDI-P, PPP, Partai Demokrasi Indonesia (Democratic Party of Indonesia, or PDI), Partai Bhinneka Tunggal Ika (Unity in Diversity Party, or PBI), and Partai Demokrasi Kasih Bangsa (Love the Nation Democratic Party, or PDKB). The latter three parties fared much better in West Kalimantan than nationwide, likely because of their *Pancasila* ideological orientation. PDI's[2] strong showing in the province in 1999 must be attributed, argues Tanasaldy (2012, 290) to the party's original components: Ikatan Pengusung Kemerdekaan Indonesia (Indonesian Independence Support League, or IPKI) and Partai Katolik (the Catholic party), that were popular in West Kalimantan in the 1970s, mostly among the Dayaks. PKB, PAN and PBB, on the other hand, all three appealing to Muslims, finished much more weakly in the province than in the rest of the country, presumably because the share of the Muslim vote in the province is much lower than in Indonesia on average.

Given that prior to the 1999 election parties had little time to target specific local electorates within provinces, it is understandable that the provincial results in West Kalimantan do not show much more than the Islamic versus *Pancasila* (or nationalist) party split, with half of all votes going to the two biggest catch-all parties: Golkar and PDI-P. The high number of non-Muslim voters resulted in a relatively poor showing of Islamic parties in West Kalimantan; especially PKB, ranked 10th in the province despite being the third most popular party nationwide, while Partai Keadilan Sejahtera (Prosperous Justice Party, or PKS), 6th in

Indonesia, came only 9th in West Kalimantan. On the other hand, Partai Nasiona-lis Banteng Kemerdekaan (Freedom Bull Nationalist Party, or PNBK, a split-off from PDI-P) came 7th in the province, being only 14th nationwide, and Partai Persatuan Daerah (nationalist Regional Unity Party or PPD) finished 11th in West Kalimantan and 23rd in Indonesia. PAN and PPP, both with a Muslim background, performed better in West Kalimantan than nationwide (Tanasaldy 2007, 296).

These results would suggest that the Muslim versus non-Muslim split was dominant in 1999. There were, however, attempts to make a more specific appeal to the Dayak category; a son of one of the former Persatuan Daya (PD) offi-cials registered a party in 1998 under the old PD logo and name, but claiming the party was nationalist, i.e. open to all (Tanasaldy 2012, 292). The initiative died within months, and in any case would not have passed the requirements of the 1999 legislation on political parties. The alternative approach was, again, to seize a nationwide party and turn its local chapter into an ethnic one. At the end of 2002 AR Mecer and other figures from Pancur Kasih joined Partai Pewarta Damai Kasih Bangsa (Messenger of Peace and Love the Nation Party, or PPDKB, a re-registered clone of an earlier party), but the party failed to gain any impor-tance nationwide. In a similar case, Hubertus Tekuwaan, son of Oevang Oeray, became PBI chairman in West Kalimantan. Because of the party's poor showing on the national level in 1999, the party was not eligible to contest in 2004. Simi-larly, PDI disappeared from the national political scene and despite its historical background as the "Dayak party", could not become the vehicle of political mobi-lization of Dayaks. Small, minority-representing parties simply could not offer what the Dayak elites were looking for: a nationwide political network whose local chapter could serve as a Dayak platform.

As the 2004 and 2009 legislative elections will show, PDI-P became the party of choice for Dayaks. As Tanasaldy notes, "the overall influence of the Dayaks in the PDI-P became more pronounced when Cornelis, the [then] Dayak regency head of Landak, was elected as the chairman at its provincial chapter. In the lead-up to the 2004 election, the majority of Dayak executive heads including the deputy gov-ernor, five regency heads and one deputy mayor had joined PDI-P" (2012, 290). As leaders needed a party to secure support for executive election nominations, as well as access to nationwide patronage through party channels, PDI-P was an almost obvious choice. The party was more oriented towards non-Muslims than Golkar, more secular and leftist, and had strong structures (unlike other small par-ties Dayaks voted for in 1999). Neither Christian parties, nor the PPD, that vied for support in the Outer Islands, could offer the same spoils of being one of the biggest players in the national politics, as PDI-P did.

The focused analysis of the 2009 and 2014 legislative election will present the parties' results according to the religious composition of sub-districts to support the statement argued here: in West Kalimantan PDI-P and Golkar have a clearly ethnically tinted image; PDI-P as an ethnic Dayak/Christian party and Golkar as a Muslim party. I argue this is due to two features: the parties' elected representa-tives and their voters' profile established based on the geographical distribution of voters. Further paragraphs, discussing gubernatorial elections, will add another feature to PDI-P and Golkar: their ethnic leadership. Note, however, that PDI-P and Golkar do not explicitly appeal to any category, and they are not ethnically exclusive parties. Interviews with PDI-P officials[3] all involved their statements that although the party is more popular among Dayaks, it is not an ethnic party; it is open to membership of all ethnic categories and it stands equally for everybody's rights and privileges. Non-Dayak representatives elected on PDI-P's ticket prove the point; so do PDI-P's Malay candidates for *bupati* and vice-*bupatis* in several elections, vying for Malays' votes by conducting prayers with Muslim imams during their rallies.

The same attitude was presented by West Kalimantan Golkar leaders, who rejected the notion of Golkar being ethnically exclusive[4], although they admitted that their party is more popular in areas with higher proportions of Malays. Golkar's governor candidate nominees and its province-level leadership also speak for the party's ethnic Malay outlook, but the claim to Golkar being an ethnic party must be seen as conditional. Among other parties, Demokrat's support will be shown to increase in urban and coastal areas, and Muslim parties expectedly enjoy higher support in constituencies with higher proportions of Muslims.

Paige Tan noted the new dynamics between parties and local strongmen created by the direct elections as early as 2006.

> As the regional election law was crafted in the party-dominated national parliament, the parties assured a monopoly role for themselves in contesting these local races. However, thus far, the dynamic appears to be somewhat different from the national picture in the regional races, a complex dance between the parties and incumbents/local notables. The party centres were allowed a say in candidate selection for the regional contests by the election law. In some cases, the parties put forth their own candidates for office from within party structure. In other cases, though, the parties have attempted to attract serving officials or those believed to have pull in the localities, due to ethnic, family or financial considerations, to run under a party banner.
>
> (2006, 96)

This point can be elaborated well with the example of West Kalimantanese executive elections. Party support is the technically easier way to run for governor or *bupati*. Those who have no party support must prove support of thousands of individual residents, which is a tedious process. Therefore, parties try to groom popular candidates within their ranks, and popular politicians attempt to build networks within party structures to secure nomination in executive elections. The need to combine on-the-ground popularity and within-party popularity accounts for personal strives and frequent party shifts. *Bupatis* who have grown popular in their areas seek recognition from their parties; that was the case of Abang Tambul Husin, *bupati* of Kapuas Hulu, who ran for provincial chair of his party, Golkar. After he lost the party contest to another popular *bupati*, Morkes Effendi from Ketapang, in 2009 (Merdeka.com 2009), Abang Tambul Husin moved to Gerindra (Gerakan Indonesia Raya, or Great Indonesia Movement), where he soon secured the position of the provincial leader (Kompas 2011). As a Gerindra nominee, he won the subsequent Kapuas Hulu *bupati* election for his second term.

Similarly, when the Demokrat party decided to back Cornelis' candidacy for governor (Cornelis' second term, 2012–2017), Demokrat's *bupati*, Sintang Milton Crosby, who wished to contest the upcoming gubernatorial election on his party ticket, quit Demokrat and joined Golkar (*Pontianak Post* 2012b). The *bupati* of Kubu Raya experienced the opposite; as a popular but non-partisan regency head, Muda Mahendrawan was approached by the Demokrat party, which was searching for a chair of its provincial branch. After the death of Henri Usman, the previous Demokrat chair in West Kalimantan, Muda Mahendrawan agreed to be the interim chair of the party, without even being the party's member. He also considered running (with his party's encouragement) for the party provincial chair position in the 2012 internal party leadership election, but declined in the end.[5]

Therefore, parties as vehicles of support in executive elections face the problem of poor candidate party loyalty. As executive heads build up their own support, which is independent from party position, party-hopping becomes endemic. However, parties' main objective in executive elections is to back the winning candidate, and not necessarily the candidate who is the closest to the party ranks. Demokrat's support for Cornelis in the 2012 gubernatorial election is just one example; Golkar's support for AM Nasir (from PPP) as a *bupati* candidate in Kapuas Hulu, against the party's own Agus Mulyana, is a similar case (although Agus later became Nasir's running mate). Similarly, in the second Sekadau election, when incumbent Simon Petrus (Demokrat) was expected to win, PDI-P supported him although the party would arguably have had little problem finding a suitable candidate among its Sekadau echelons. Hence, in most cases the ethnic image of parties is lost in personal bickering and individual popularities.

Political parties can also be looked at according to their success rates in *bupati* candidate endorsement if the party is not eligible to nominate on its own. Within Indonesia, Golkar is the sole winner in terms of elected executive offices filled by its candidates without forming a coalition, while PKS and PAN are frontrunners in terms of winning *bupati* posts as members of supporting coalitions (Maulana and Situngkir 2009, fig. 4). While Golkar's individual performance is explainable by its excellent showing in legislative elections, which entitles the party to nominate *bupati* and governor candidates on its own, one cannot fail to recognize PKS's and PAN's ability to not only identify the potential winning candidate, but also maintain such a position on the party scene that makes them attractive coalition parties. PPP is another party that succeeds in standing behind the right contenders. A good example of PPP's strategy in executive elections was the case of Hasan Karman (member of Partai Indonesia Baru, or New Indonesia Party) in the 2007 Singkawang mayor election (*Harian Equator* 2007). PPP lent its support to the only Chinese candidate in the race, despite the party being a Muslim one. It was clearly more important to the party to support the winner than to support a Muslim.

Interestingly, PDI-P was quite consistent in supporting Christian Dayak candidates in those regencies/cities where this category could influence the outcome of the election. Especially in Ketapang and Kapuas Hulu (both first and second direct elections), PDI-P opted for a Dayak candidate in polls bound to be split between Malays and Dayaks. One exception was the 2010 Melawi election, where PDI-P supported a Malay with a Dayak running mate, although Melawi's ethnic proportions are similar to those of Ketapang and Kapuas Hulu and Dayak could be a winning category on its own. PDI-P backed tickets with Dayaks as running mates in Singkawang (first election) and Pontianak regency (first election), despite the very low numbers of Dayaks in these areas that make this category only marginally relevant there. Nevertheless, in both cases, sub-districts with high numbers of Dayaks indeed voted for the pairs with Dayak running mates.

Other parties' executive candidate nominations were less obviously ethnically informed: although Golkar was in general more likely to back a Malay candidate (especially in its strongholds of Ketapang and Kapuas Hulu), there are no cases of the party supporting a Malay underdog (compare PDI-P and Dayaks in Singkawang and Pontianak regency) just to send an ethnic message. Note that Golkar and PDI-P are the two frontrunners for popularity in West Kalimantan, and hence staunch enemies. The two parties have so far only once (in a total of 30 direct executive elections held in the province, including the gubernatorial polls) entered a coalition together.[6] Hence, because PDI-P continues to support Dayaks, Golkar logically aims to attract the opposite of the ethnic spectrum, which in this case are Malays. Note that PDI-P and Golkar entered coalitions together on the national level (e.g. Golkar officially supported Megawati's presidential candidature in 2004), as well as in other provinces, towns or regencies across the country.

In West Kalimantan, Golkar was most likely to go an election alone, or to enter a coalition with one of the Islamic parties (e.g. PKS in Sambas 2011 and Ketapang 2010). Demokrat and PDI-P were found supporting common candidates on many occasions, often with success (vide 2012 gubernatorial election, Ketapang, Sintang and Sekadau 2010 *bupati* elections). Note that this is in contrast to the national-level politics, where PDI-P continuously stayed out of the ruling coalition led by the Demokrat chief and president Yudhoyono throughout the 2004–2014 period (Mietzner 2013, chap. 5). The more in-depth analysis of parties' motivation in entering particular coalitions in executive elections is beyond the scope of this research, however, the general picture suggests that most of these coalitions are marriages of convenience, or attempts to bet on the winner and obtain the best cost-benefit ratio. No evidence so far suggests that parties would not cooperate with partners because of ideological differences.

To conclude, parties rather tend to lend support to a candidate that is likely to win (especially if it is an incumbent), than to field candidates from their own ranks who might have a lesser chance of winning. A party's ethnic image and its supported candidate's ethnicity were rarely consistent, as the foregoing examples showed, and the only party that displayed some ethnic consistency in its executive candidate nomination was PDI-P.

5.2 Legislative elections

The formal elements of the electoral process were presented in Chapter 4. From the institutional canvas (i.e. pure PR electoral system), combined with the high number of competing parties, the analysis of individual candidates' electoral performance and matching it with ethnic preferences becomes unfeasible. Relatively small constituencies on the third tier of administration inform that many candidates will be directly, not merely through media, known to their voters. In this case, assuming that a voter is most likely to vote for a person they know best, we should expect voters to opt for those candidates who are geographically closest to them. This way we do assume that a local ethnic category is activated, but it cannot be proven within this research. The provincial and national legislative representatives are elected in much bigger constituencies and it is likely that the candidates are only known to the voters through media. Although geographical proximity may still be a valid incentive, and so a local ethnic category may be activated, these two elections may again be cleaved along religious lines, or the Dayak-Malay-Chinese-Madurese line, given that these categories can be ascertained from names, outfits, Election Commission information and press releases about candidates (compare in Appendix 1). I offer, however, no evidence of these statements, and these are merely informed assumptions. In all, the analysis of the electoral result will focus on preferences for parties expressed in these elections to see how parties' electoral results reflect ethnic divisions in the province.

In 2004 Golkar dominated the elections in West Kalimantan by obtaining more than 24% of the votes, while PDI-P received only 17% and PPP 8%. The fourth party, Demokrat, collected 6% of the votes; eight other parties won seats (between one and four) in the provincial DPRD.[7] Among the elected DPR representatives, there were seven Malays (elected on the tickets of PPP, Golkar, PAN, Partai Bintang Reformasi (Reformed Star Party, or PBR), and PDI-P), two Dayaks (fielded by PDI-P and PDS) and one Chinese (from the Demokrat list). The elected DPD members from West Kalimantan in 2004 were two Dayaks and two Muslims (none of the Muslims was of West Kalimantan origin).

In the 2009 general election, PDI-P in West Kalimantan performed remarkably well on the DPR tier and 3 out of the total of 10 seats assigned to the province in the parliament went to this party. Golkar had two representatives from West Kalimantan and so did Demokrat. Three other parties (PKS, PAN and PPP) saw their representatives elected to DPR from West Kalimantan. PDI-P's stunning result in the election to DPR must be attributed to the personal success of Karolin Margret Natasa, the daughter of West Kalimantan governor Cornelis (Cornelis is also PDI-P's provincial chief). Karolin alone obtained 47% of all votes cast for PDI-P in West Kalimantan in the election to the national parliament.[8] In the election to the provincial DPRD and regency/municipality assemblies, PDI-P's result was lower than that to DPR by precisely the number of votes that Karolin Margret Natasa obtained. The ethnic split of the elected representatives in that election is well worth noticing; PKS's, PAN's, Golkar's and PPP's elected MPs are all Muslims; except for Rahman Amin, who hails from Jakarta and is not a

West Kalimantanese, the other four Muslims are Malays. All three PDI-P MPs are Christian Dayaks. Demokrat's representatives from West Kalimantan are both Chinese.

Figures 5.1 to 5.3 reflect the biggest parties' electoral results in 2009 in *kabupaten* compared with the religious composition. DPR-level data is used for this analysis. Figure 5.1 shows what seems like a positive – albeit highly inconsistent – correlation between Muslims and support for Demokrat. I argue, however, that Demokrat popularity was higher in the coastal *kabupaten* and municipalities[9]: Sambas, Singkawang, Bengkayang, Pontianak regency and city, Kubu Raya and Ketapang. In these areas, numbers of Muslims are higher on average, hence the rising trend line, but the preference for the party was not necessarily only by Muslims. Detailed data showing the party's performance by sub-district (*kecamatan*) was not available, so the proof is weak, but along the coast the average support for Demokrat was 15%, while in the interior regencies Demokrat's result was only 7%

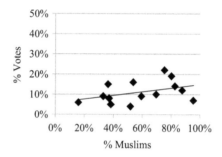

Figure 5.1 2009 electoral performance of Demokrat (DPR level) against the proportion of Muslims in regencies in West Kalimantan

Source: Author's own compilation according to Badan Pusat Statistik (2010) and Komisi Pemilihan Umum (2009a).

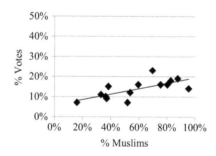

Figure 5.2 2009 electoral performance of Golkar (DPR level) against the proportion of Muslims in regencies in West Kalimantan

Source: Author's own compilation according to Badan Pusat Statistik (2010) and Komisi Pemilihan Umum (2009a).

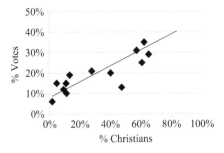

Figure 5.3 2009 electoral performance of PDI-P (DPR level) against the proportion of Christians in regencies in West Kalimantan

Source: Author's own compilation according to Badan Pusat Statistik (2010) and Komisi Pemilihan Umum (2009a).

Note: There was no difference between Protestants' and Catholics' preferences for PDI-P.

on average. A commonsensical explanation of Demokrat's popularity on the coast would be quite simple: Demokrat is a new party without established structures at the sub-district level and below. The West Kalimantanese overwhelmingly rural interior is difficult to access and the rural population has few ways of acquainting itself with new party organizations. The coastal areas are more urbanized and are susceptible to new players in politics, hence their higher support for Demokrat.

By contrast, Figure 5.2 shows the more true-positive correlation between the Muslim population and Golkar's support. The distribution of points along the trend line is much more consistent. Golkar's performance did not vary according to regional differences in the province.

Figure 5.3 clearly indicates that PDI-P is most popular among the Christian voters. This is consistent with the party's nationwide image as a party of Christians, minorities or non-Muslims.

Among other parties, PKS had the strongest positive correlation with Muslim proportions, while PDS was more popular in sub-districts that have higher numbers of Christians. Gerindra's, Hanura's, PPP's and PAN's results were independent of religious composition; however, this analysis is based on electoral results tabulated by regency. Between the three big parties (Golkar, PDI-P and Demokrat), only Demokrat increased its result in 2009 compared to the 2004 election (from 7% to 12%) in West Kalimantan. Both Golkar and PDI-P's support dropped between the two elections (from 24% to 14% for Golkar and from 17% to 15% for PDI-P), but the decrease in support for the two parties is partly explained by the higher number of competing parties (44) in the 2009 election (in 2004 only 24 parties qualified to compete). Golkar's drop was much sharper than PDI-P's and the net change in the parties' popularity indicates that Golkar indeed lost supporters, while PDI-P maintained its popularity.[10]

The 2009 election also showed that popularity of a single candidate, like that of Karolin Margret Natasa, can pull the entire party's (in this case PDI-P) result by enormous margins. Karolin's electoral performance has to be further attributed to her father's – Cornelis, the two-time governor of West Kalimantan – popularity in the province. Hence, PDI-P is the governor's party in West Kalimantan, much like Demokrat was the president's party on the national level. A similar pattern is visible in PPD's win in the Kayong Utara regency. Two popular politicians (one of whom is the regent) from the regency are leaders of the party, and PPD obtained 45% of the votes in the area.

The 2014 general election showed the same trend among parties in West Kali-mantan. PDI-P won in the province with 33% of the votes, much higher than nationwide (19%). Karolin Margaret Natasa this time received the highest number of votes of all candidates across the country, almost 400,000; of those 128,000 came from the single regency of Landak, where her father used to be the regent. Golkar came second (both in West Kalimantan and nationwide), and the party's performance in the province was similar to its overall result. Gerindra came third and Demokrat fourth, corresponding to their position nationwide, but both par-ties received a slightly smaller proportion of votes in the province than across the country. PDI-P in West Kalimantan won three seats in the national parliament (all three representatives are Dayaks and Catholics), the seven other candidates each came from different parties; Demokrat's representative is a Protestant Dayak and Gerindra's is a Catholic Dayak, while Golkar's, PAN's, PPP's and Demokrat's representatives are all Malays. PKB's elected representative from the province is a Jakarta-born Chinese and Buddhist.

PDI-P's support again came from Christians (Figure 5.4), while Golkar's came from Muslims (Figure 5.5), and these results are not only consistent with the pre-vious elections, but also show that both parties have entrenched their respective ethnic positions and have consolidated their support from the target ethnic cat-egories (compare Figures 5.3 and 5.2). Gerindra's (Figure 5.6) and Demokrat's (Figure 5.7) respective popularity were independent of religion. In the 2014 elec-tion, Demokrat displayed about even popularity across regencies, showing that it

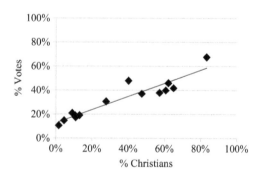

Figure 5.4 2014 electoral performance of PDI-P (DPR level) against the proportion of Christians in regencies in West Kalimantan

Source: Author's own compilation according to Badan Pusat Statistik (2010) and Komisi Pemilihan Umum (2014).

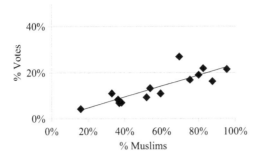

Figure 5.5 2014 electoral performance of Golkar (DPR level) against the proportion of Muslims in regencies in West Kalimantan

Source: Author's own compilation according to Badan Pusat Statistik (2010) and Komisi Pemilihan Umum (2014).

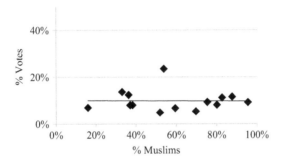

Figure 5.6 2014 electoral performance of Gerindra (DPR level) against the proportion of Muslims in regencies in West Kalimantan

Source: Author's own compilation according to Badan Pusat Statistik (2010) and Komisi Pemilihan Umum (2014).

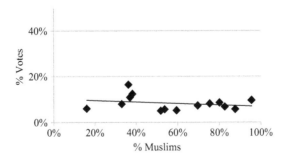

Figure 5.7 2014 electoral performance of Demokrat (DPR level) against the proportion of Muslims in regencies in West Kalimantan

Source: Author's own compilation according to Badan Pusat Statistik (2010) and Komisi Pemilihan Umum (2014).

succeeded in penetrating the interior of the province. Eight out of 12 contesting parties will have representatives from the province in the national parliament.

As in the previous elections, PDI-P's performance on the provincial tier was weaker and the party obtained 23% of the votes for the provincial legislature (compared to 33% for the DPR level). On this tier, Golkar and Demokrat both received almost 12% of the votes, which for Golkar meant a worse result than on the DPR level, while for Demokrat it was a much better performance. Gerindra received only 9% of votes on the provincial tier. One party (PBB) out of 12 contesting did not win a seat in the provincial DPRD in Kalimantan Barat.

5.3 Presidential elections

Presidential elections might seem a poor indicator for ethnic preference on a regional level, as invariably all presidential candidates in Indonesia are Muslims and most of them are Javanese; moreover, political divisions in the regions are different from those on the national level or in Java. Nevertheless, this research is committed to establish whether direct elections on the highest administrative level can provide an environment that activates ethnic categories alternative to those activated in elections on other levels. Therefore, we will test whether a direct presidential election produces an environment that results in redefinition of minimum-winning coalitions within the province for this election, as hypothesis H1 presented in Chapter 1 suggests.

One potential result of the presidential election would be the emergence of a provincial (e.g. West Kalimantanese) identity or an "Outer Island" ethnic loyalty (i.e. non-Javanese identity or centre-periphery cleavage). The 2004 presidential election was a good test for these potential categories' activation. In this election, the PPP presidential candidate and one of the party's leaders, Hamzah Haz, was a Malay from West Kalimantan, born in Ketapang (F. Alkap Pasti 2003, 141). Hamzah Haz was also the vice-president during Megawati Sukarnoputri's presidency (2001–2004). If the provincial West Kalimantanese identity were activated during this election, we would have seen increased popularity of this candidate, and not only among Malays and Muslims, but also among the Christians or Chinese in the province. Notably, Hamzah's ethnic background or origin did not play any significant role in the campaign, but, arguably, if he had been perceived as having any potential to win, West Kalimantanese voters might support him hoping to gain some extra leverage from the common origin. Hamzah Haz did not appear to be a hopeful contender and his performance in the province was abysmal. Even in his home *kabupaten* Hamzah Haz finished only fourth, while within the entire province he came fifth (out of five candidates). Hence, we know that the regional identity ("West Kalimantanese") was not activated during that presidential election, and, similarly to examples from Sarawak, voters showed that they prefer to vote for a winner than to vote for their co-ethnic, if the latter has no chance of winning.

Studying the 2004 contest for the president's office, we also see that the Java/Outer Island cleavage was not important in West Kalimantan. In 2004 there were two pairs of candidates including non-Javanese: Susilo Bambang

Yudhoyono-Yusuf Kalla (the second from Sulawesi) and Hamzah Haz-Agum Gumelar (Javanese). The election result shows, however, that the voters were not deterred by Javaneseness of other candidates, and Yudhoyono's good showing has to be attributed to factors other than Kalla's Outer Island background.

Hence, the two commonsensical cleavages to be observed on the regional level ("West-Kalimantanism" and centre vs. periphery) did not present in the studied province. To the contrary, in both the 2004 and 2009 elections the contest – both nationwide and in West Kalimantan – turned out to be between Megawati Sukarnoputri (paired with Hasyim Muzadi, both Javanese) of PDI-P and Susilo Bambang Yudhoyono (with Yusuf Kalla) of Demokrat.

The Megawati–Hasyim Muzadi pair won in West Kalimantan thanks to the Christian/Dayak votes (compare Figure 5.8), although by the narrowest of margins (50.02% against the 49.82% Yudhoyono obtained). The total percentage of non-Muslims in the province is about 43% and the percentage of Christians is 32%; hence to achieve the said result the PDI-P–supported pair must have attracted some Muslim voters. In general, the Muslim vote was split in the first round, but mostly consolidated around the pair Susilo Bambang Yudhoyono–Yusuf Kalla in the second round.

During the 2009 presidential campaign Megawati (paired this time with Prabowo Subianto) visited West Kalimantan twice (*Pontianak Post* 2009a, 2009b). This time, however, her performance in the province was weaker; the Megawati-Prabowo team obtained 37% of the votes (against their nationwide result of 27%). Christians in the province remained Megawati's most loyal voters (see Figure 5.9).

Although Yudhoyono's (paired with Boediono) electoral result in West Kalimantan (54%) was lower than the average in the country (61%), this time he won comfortably in the province. Yudhoyono scored 70% or higher in four coastal/Malay/urban administrative regencies of the province (Pontianak city, Pontianak regency, Kubu Raya regency and Singkawang city), which is consistent with his party performance in the legislative election that year. Megawati's popularity in

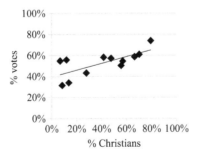

Figure 5.8 Megawati–Hasyim Muzadi electoral result according to the Christians' proportions by regency in the 2004 presidential election (second round) in West Kalimantan

Source: Author's own compilation according to Kementerian Agama (2010) and Komisi Pemilihan Umum (2004).

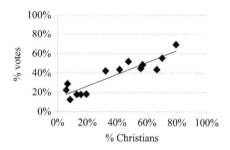

Figure 5.9 Christian proportions in regencies and Megawati-Prabowo's electoral result in the 2009 presidential election in West Kalimantan

Source: Author's own compilation according to Kementerian Agama (2010) and Komisi Pemilihan Umum (2009b).

West Kalimantan can be explained by the strong local standing of her party, PDI-P. The two visits to the province during the campaign periods indicated Megawati's special attention for her West Kalimantanese voters, who were shown here to be Christians/Dayaks. Therefore, the presidential election's results add to the argument of PDI-P being an ethnic party. PDI-P's voters identify with the party even in elections in which the party's candidate is of a different ethnic background than that of the voters.

The 2014 presidential election in Indonesia was a very different matter, and in many ways it was about much more fundamental issues than just a choice between two pairs of candidates. For the first time, there were only two contesting pairs of candidates, and consequently only one round. As Mietzner (2014) showed, Indonesians were voting for or against upholding democratic principles and were choosing between elitist politics rooted in the Suharto era and fresh-blood, commoners-related political force. Prabowo Subianto from Gerindra, former son-in-law of Suharto and an army general, running with Hatta Rajasa from PAN, competed against a political newcomer, Joko Widodo, the Surakarta (Solo) mayor from 2005 to 2012, and the Jakarta mayor from 2012 to 2014. Jokowi, as he is commonly known, was supported by PDI-P, and Jusuf Kalla (also abbreviated as JK) was his running mate, despite being a Golkar member and Golkar officially backing the other presidential pair.[11] The results, cross-checked against the religious proportions in provinces across Indonesia, show the expected preference of non-Muslims towards Jokowi, but nationwide, religion does not explain the voters' behaviour entirely. Although support for Prabowo–Hatta Rajasa and Islam adherence are minimally correlated, there are clearly other factors in play influencing voters' preferences (see Figure 5.10).

Despite the quite dramatic outlook of the campaign on the national level, in West Kalimantan the 2014 election was not much different from the previous presidential contests. PDI-P was as strong in the province as before and the party's nominee had an upper hand. Just like Megawati in 2009, Jokowi personally campaigned in West Kalimantan (Kompas 2014), and immediately upon his arrival

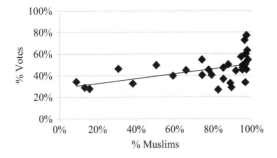

Figure 5.10 Support for Prabowo–Hatta Rajasa according to Muslim proportions in Indonesian provinces in the 2014 presidential elections

Source: Author's own compilation based on Badan Pusat Statistik (2010) and *Tribun Pontianak* (2014).

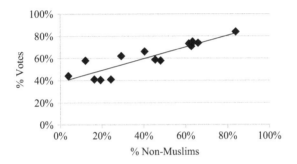

Figure 5.11 Support for Jokowi-JK according to the numbers of non-Muslim voters in West Kalimantan regencies in the 2014 presidential election

Source: Author's own compilation based on Kementerian Agama (2010) and *Tribun Pontianak* (2014).

in Pontianak he clad himself in traditional Dayak headgear. Megawati also visited the province during the campaign, and Governor Cornelis was the head of Jokowi's campaign team in the province. Jokowi-JK won comfortably in West Kalimantan, obtaining 60% of the total vote. As Figure 5.11 shows, the vote was strongly dependent on voters' religion, and, although many Muslims voted for Jokowi, the gross of support for him came from non-Muslims.

In the previous paragraphs, I showed that local West Kalimantanese party preferences (and ethnic preferences) are powerful enough to decide the fate of presidential candidates in the province. The results presented earlier show that the presidential election is hijacked by the provincially valid cleavage. There is no indication of voters in West Kalimantan in presidential elections activating any other category than the ones known to be activated at the provincial level and through party loyalty. Especially one party's (PDI-P) ethnic ties are strong enough to cleave consecutive presidential elections.

5.4 Gubernatorial elections

The introduction of direct elections for governor and regency heads in Indonesia created a new set of incentives for ethnic mobilization. The electoral result in legislative elections remains important because of the necessary party support for gubernatorial (as well as mayoral and *bupati*) candidates, but party loyalty ceased to be the decisive factor in governor selection. The 30% threshold in the first round makes the situation particularly interesting: the vote can be split between as many as three tickets and mathematically all of them can achieve the required margin in the first round. The 2007 election was the first direct gubernatorial election in West Kalimantan. The incumbent pair contested along with three other pairs of candidates.

Figure 5.12 presents the candidates' ethnic backgrounds. Usman Ja'afar and Laurentius Herman (LH) Kadir were the incumbent governor and vice-governor; part of their campaign was devoted to the separation of the five eastern *kabupaten* in West Kalimantan and to the creation of a new province called Kapuas Raya. This was likely to resonate well with the residents in the five eastern *kabupaten*, as it would add importance to their region. The strategies followed by candidates with ballot numbers 1, 2 and 3 were clearly similar: all pairs included a Malay and a Dayak, or a Muslim and a Catholic, each pair combining different regions of the province, supported by one big party (1 and 2) and/or a coalition of small parties (1, 2, 3). The candidates' profiles followed the logic of the religious proportion of the West Kalimantan population, with Muslims as the majority and Christians as a significant runner-up. All candidates except AR Mecer made an extensive use of ethnically related attributes (compare their official images presented by the Election Commission in Appendix.). The Malay candidates wore the traditional black headgear. The non-Muslim candidates chose attributes reflecting their Dayak background (colourful, beaded headgear and/or jackets). Christiandy Sanjaya, a Chinese, appeared clad in a jacket resembling traditional Chinese outfits (collar, buttons). Judging by the attributes, the Malay/Muslim candidates were appealing to "Muslim" voters, while non-Muslim candidates chose to reflect their "Dayak" or "Chinese" identities in their attire. There is, however, no specific "Christian" outfit that one could choose, and religion is already included in official materials about candidates prepared by the Election Commission.

Candidates 1, 2 and 3 were visibly more inclusive in their strategy: the Muslim and Christian religious background could appeal to about 90% of the West Kalimantan population. Also as Malay-Dayak, all three pairs were inclusive, and regionally each of them included candidates from the coast and the interior of the province. On the other hand, ballot number 4 candidates' combined ethnic membership could be shared by a much smaller number of West Kalimantanese. Cornelis and Christiandy are a Catholic/Protestant pair and as such Malays, Madurese and Javanese share no ethnic categories with them. If seen as a Dayak-Chinese team, they include Muslim Dayaks, but exclude all other Muslims nevertheless. No other Dayaks-Christians were running as gubernatorial candidates (other Dayaks were running mates); moreover, there were no other Chinese candidates, and Christiandy, as the sole Protestant, could hope to attract the entire Protestant vote.

Table 5.1 The 2007 West Kalimantan governor candidates' profiles

	Ticket 1	Ticket 2	Ticket 3	Ticket 4
Name	Usman Ja'afar / Laurentius Herman Kadir	Oesman Sapta Oedang / Ignatius Lyong	Akil Mokhtar / Anselmus Robertus Mecer	Cornelis / Christiandy Sanjaya
Religion	Malay / Dayak Catholic	Malay (or Bugis)[i] / Dayak Catholic	Malay / Dayak Catholic	Dayak Catholic / Chinese Protestant
Region of origin	Sekadau / Putussibau/ Kapuas Hulu	Sukadana/ Ketapang[ii] / Putussibau	Putussibau/ Kapuas Hulu / Ketapang	Sanggau / Singkawang

Source: See Appendix 2.

i Tanasaldy (2012, 269) holds Oesman for Malay (2012, 269), but elsewhere Oesman was identified as of Bugis origin (Lingkaran Survei Indonesia 2008, 4).
ii Sukadana later became the capital of a new *kabupaten*, Kayong Utara.

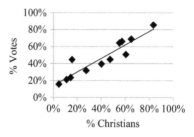

Figure 5.12 Support for Cornelis-Christiandy by regency and proportions of Christians in the 2007 West Kalimantan gubernatorial election

Source: Author's own compilation according to Badan Pusat Statistik (2010) and Komisi Pemilihan Umum Provinsi Kalimantan Barat (2007).

Cornelis-Christiandy also sought to attract Madurese (Muslims), despite not sharing membership in any ethnic category by the candidates and the Madurese voters, except both Madurese and Chinese being *pendatang* (i.e. incomers). Cornelis-Christiandy "used the issue of returning the Madurese to Sambas as one of their strategies to get support and sympathy from ethnic Madurese" (CAIREAU Center for Acceleration of Inter-Religious and Ethnic Understanding 2008, 2). The Madurese had fled the Sambas region because of the violent conflict with the indigenous Malays there in 1999. The head of a Madurese non-governmental organization in West Kalimantan (Ikatan Keluarga Besar Madura, or Association of the Great Madurese Families) in a 2012 interview claimed that Madurese support for Cornelis-Christiandy in 2007 was at 40%.[12] There is no evidence, however, to substantiate this assertion.

Cornelis-Christiandy won handsomely in the first round with 43.67% of the votes. Usman Ja'afar-LH Kadir came second with 30.94% of the votes, Oesman Sapta-Lyong placed third with 15.74% of the votes, and Mokhtar-Mecer obtained only 9.66% of the votes. Figure 5.12 shows that Cornelis-Christiandy's support was strongly correlated with numbers of Christians. Both Catholics and Protestants voted uniformly for this pair (graphs not shown here). The religious data is of no use to establish whether Madurese and Javanese voted for Malay or non-Malay candidates, but the exact alignment of the non-Muslim vote with support for Cornelis-Christiandy suggests that all Muslim voters, including Madurese and Javanese, voted for another pair. Figure 5.13 accounts for the Chinese vote, as ethnic Chinese are all Buddhists or Confucianists or Christians, and shows that virtually no Muslim vote was cast for the Dayak-Chinese pair.

Were regional ethnic categories activated in this election? Cornelis' support was dependent only on non-Muslim voters, without any discernible differences between regions in the province. Oesman Sapta-Ignatius Lyong won in Oesman's home *kabupaten*, Ketapang; however, the support was concentrated in the relatively small area of Sukadana (the five sub-sub-districts that later became a separate *kabupaten*, Kayong Utara, and five other *kecamatan* neighbouring this area). Akil Mokhtar's performance was strongest in Kapuas Hulu where he hails from,

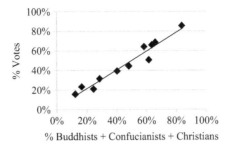

Figure 5.13 Support for Cornelis-Christiandy by regency and proportion of non-Muslims in the 2007 West Kalimantan gubernatorial election

Source: Author's own compilation according to Badan Pusat Statistik (2010) and Komisi Pemilihan Umum Provinsi Kalimantan Barat (2007).

but not strong enough to win against Cornelis-Christiandy in this regency. AR Mecer's home *kecamatan* in Ketapang, Sungai Hulu, was strongly behind him, but in Ketapang as a whole Mokhtar-Mecer's performance was actually lower than the average in the province. Cornelis-Christiandy won in all five eastern *kabupaten*, although this can hardly be attributed to regionalism. Although Cornelis hails from Sanggau, for most of his career he was active in Landak and Bengkayang, and these are the areas with which he is commonly associated, also because of his membership in the Kenayatn sub-category among Dayaks (Kenayatn are concentrated in Landak and Bengkayang). There is, however, no indication that Kenayatn voted for Cornelis any more than other Dayak sub-categories. Usman Ja'afar and Ignatius Lyong won by wide margins in Pontianak city, Pontianak *kabupaten* and Sambas, although neither of them is from these areas. Their popularity there has to be attributed to Muslim or Malay loyalties. Ja'afar-Lyong attempted to mobilize the eastern regencies around the issue of creation of the new province of Kapuas Raya, but to no avail.

To sum up, the first gubernatorial election was aligned along the religious cleavage (non-Muslims vs. Muslims). I showed that the Chinese vote was consistent with the Christian vote. The Chinese vote was likely to be attracted by the Chinese running mate of the Cornelis-Christiandy ticket. It was shown, however, that the Dayak-Christian voters were not attracted to the Dayak-Christian running mates of Malay-Muslim candidates. The only Protestant candidate in this election was paired with a Catholic, and the entire Christian vote went to this pair. Moreover, there were no Madurese or Javanese candidates. There was an attempt to mobilize the Madurese to vote against Malays, i.e. not to follow religious loyalty. The result of this attempt is unknown, but the analysis presented here indicates that virtually *no* Muslims (i.e. also Madurese) voted for Cornelis-Christiandy. Moreover, regional preferences, except for a few localized outliers, were not significant. Hence, the election was a single cleavage affair, and the voters' behaviour was stunningly dichotomous. The Chinese and Dayaks joined ranks and won against Muslims, whose vote, albeit split between three Muslim candidates, was entirely against the Christian pair.

The 2012 elections brought some changes, despite being seemingly a similar affair of Dayak-Chinese (Christians) competing against Malays (Muslims). Cornelis' run for the second term was beyond doubt and the Demokrat party entered a coalition with PDI-P to support him. Critically, Demokrat presented three potential vice-governor candidates to Cornelis for consideration (*Tribun Pontianak* 2012a): Christiandy Sanjaya (Chinese/Protestant), Suryadman Gidot (Dayak/Catholic, sitting Bengkayang *bupati*) and Paryadi (Madurese, sitting vice-mayor of Pontianak). Note that a coalition with a Malay was not considered. Each of the proposed alternative running mates would complement the coalition to make it a minimum-winning one, either by the king-maker position of a Madurese or Chinese, or by making the coalition entirely Catholic Dayak. Cornelis chose Christiandy as his running mate, making it the only election among all executive head elections in West Kalimantan that the incumbent pair ran again together for the second term.

Several elements could have made the 2012 election more difficult for Cornelis-Christiandy than the 2007. Firstly, another popular Dayak/Christian candidate could split the electorate. Especially Dayak *bupatis* had been growing powerful and ambitious in their respective *kabupaten* and could threaten Cornelis' position as the Dayak leader. The second difficulty was that the Muslim elites would this time manage to unite around one strong Muslim-exclusive candidate pair, which would result in a one-on-one Christian versus Muslim election. Each of these two possibilities could lead to a win for the Muslim candidate in the first round, or take the election into a run-off that would most likely also see the Muslim pair win. In the end, however, neither was there a serious Dayak contender, nor did Malays manage to unite around one candidate.

Three (out of the total four) Malays in the contest were former *bupatis*. Two of them ran together: Morkes Effendi from Ketapang with Burhanuddin A. Rasyid from Sambas (both Malays), while Abang Tambul Husin, former *bupati* of Kapuas Hulu paired with Barnabas Simin (a local Dayak leader from Kapuas Hulu). The fourth pair included Armyn Ali, an army officer, and Fathan A. Rasyid, brother of Burhanuddin A. Rasyid. The last pair was the most surprising, as Armyn Ali, in order to run for the office, had to quit his high army post with no opportunity to return and before reaching retirement age (Antara 2012), while Fathan was running against his brother. Table 5.2 shows the profile of all the candidates in the 2012 gubernatorial election. Among the challengers, Morkes Effendi-Burhanuddin were the frontrunners, being nominated by Golkar and having been popular *bupatis*. Tambul Husin's candidacy was not unexpected, but having failed to receive backing from any of the big parties, and his support being concentrated in a single, albeit large, remote regency, he was bound to lose.

Arguably, the biggest threat to Cornelis-Christiandy domination of the non-Muslim vote came from Milton Crosby. Crosby, the sitting *bupati* of Sintang and a Demokrat member, hoped for an endorsement from his party as a gubernatorial candidate, but Demokrat decided to enter a coalition with PDI-P and support Cornelis. Milton Crosby subsequently withdrew his membership from Demokrat and joined Golkar, also taking up the function of coordinator of volunteers for Effendi-Burhanuddin candidates in the eastern region of West Kalimantan (*Pontianak Post* 2012b).

Table 5.2 The 2012 West Kalimantan governor candidates' profiles

	Ticket 1		Ticket 2		Ticket 3		Ticket 4	
Name	Cornelis	Christiandy Sanjaya	Armyn Ali Anyang	Fathan A. Rasyid	Morkes Effendi	Burhanuddin A. Rasyid	Abang Tambul Husin	Barnabas Simin
Background	Dayak Catholic	Chinese Protestant	Malay	Malay	Malay	Malay	Malay[i]	Dayak Protestant
Region of origin	Sanggau	Singkawang	Singkawang	Sambas	Ketapang	Sambas	Kapuas Hulu	Landak

Source: See Appendix 3.

i According to Benny Subianto, Abang Tambul Husin is a Muslim Dayak (2009, 337). Others describe him as Malay (Eilenberg 2012, 277), which is in line with the general understanding in West Kalimantan that being an indigenous Muslim is a sufficient criterion to claim membership in the Malay category.

Milton's move to support the Effendi-Burhanuddin candidacy added strong Kapuas Raya overtones to the campaign. Promotional materials of the Golkar candidates read, "United [let's] realize Kapuas Raya Province" (*Pontianak Post* 2012c). Morkes Effendi (former *bupati* of Ketapang) and Burhanuddin (former *bupati* of Sambas) could originally count mostly on the voters in the western flank of West Kalimantan. With Milton Crosby on board, and in particular with his regionalism, the Golkar candidates had a chance to spread into the eastern *kabupaten*.

Milton's move can be explained by tension between him and the governor pertaining to the process of formation of the new province out of the eastern part of West Kalimantan. Milton Crosby had been the coordinator of the process leading to the establishment of Kapuas Raya Province, and he was the strongest proponent of the provincial division. Cornelis, on the other hand, has shown what can be interpreted as a lack of commitment to the cause of Kapuas Raya. In June 2012, Cornelis argued that the creation of the new province should happen only around 2020 (much later than the province's proponents hope for), to make sure that the area is well prepared to be a separate unit (*Tribun Pontianak* 2012c). The ethnic proportions of the two provinces after the split may explain Cornelis' lukewarm attitude to Kapuas Raya. Critically, the split of the province would render West Kalimantan a clear Muslim-majority unit (about a 2:1 proportion of Muslims to non-Muslims) and the Dayak position in the politics of this province would diminish. Kapuas Raya, on the other hand, would be a Christian-majority province, albeit with a substantial (about 40%) Muslim component. Cornelis' strongholds, Landak and Bengkayang, would remain part of West Kalimantan. These facts also suggest why Effendi and Burhanuddin (both from the west part of the province) could well afford the coalition with Milton – by parting with the five eastern regencies, they would not lose any of their supporters, as they have few voters in the Kapuas Raya region. After the formation of Kapuas Raya, their position in the West Kalimantan province would not be threatened.

While officially declaring that his candidacy seeks the support of all ethnic categories in West Kalimantan, in his promotional materials Cornelis emphasized his membership in the Dayak category, never failing to sport Dayak embroidery on his red jacket. However, the Chinese and Madurese support was also prominent in his campaign. Christiandy, his deputy, wore blue for his party affiliation (Demokrat), but invariably with Chinese-style finish (collar, buttons). Yakobus Kumis, campaign team member and an important Dayak figure, claimed that a "Dayak-Chinese-Madurese" coalition supported the Cornelis-Christiandy pair[13] and indeed the Madurese IKBM (Ikatan Keluarga Besar Madura, or Association of Madurese Great Families) head had been touring the province to convince Madurese to vote for Cornelis and Christiandy. Among Cornelis' supporting parties was also Partai Kebangkitan Bangsa (PKB), related to the Muslim Nahdlatul Ulama (NU) organization. PKB and NU provided Muslim *ulamas* for prayers and to reassure voters that it is an election of the province political leader, and not a religious leader, and there is no religious obligation on Muslims to vote for a Muslim.[14]

Supporters of PKB could help fight the notion some Islamic teachers spread that Muslims are not to vote for non-Muslim candidates, or, in a wider sense, Muslims should not be led by non-Muslims. Similarly, IKBM was shown to contribute to Cornelis' campaign through the presence of its chairperson, Sarumli Seneh, at the rallies. He expected that the vast majority of the Madurese community would vote for Cornelis.[15] There is no data that corroborate these claims. The Madurese leaders involved in Cornelis' campaign emphasized that the reason they support him is his peaceful administration of the province for the previous five years.[16] The propaganda that depicted Cornelis as a peacemaker may have resonated well with Madurese, many of whom had been driven away from their homes in Sambas. The running mate of the strongest Malay candidate pair, Burhanuddin A. Rasyid, had been the *bupati* of Sambas during the Malay–Madurese clashes, and, as my interviewee underscored[17] the Madurese still associated him with failure to maintain peace.

Tambul Husin (with Barnabas) focused on maintaining their regionally concentrated support (*Rakyat Kalbar* 2012), and arguably, Milton Crosby's attempt to steer Morkes' campaign towards the Kapuas Raya issue only sharpened Tambul Husin's commitment to the provincial split cause. Tambul announced that "only a person from the interior [*hulu*] knows [the] desires of the people in the Eastern region of Kalimantan Barat" (*Rakyat Kalbar* 2012). This strong regionalism was not particularly new, as in the previous election Ja'afar-Lyong raised the issue of Kapuas Raya, but the prominence it gained in the 2012 election suggests an attempt to shift voters' preference away from religion towards region, accounting for new category activation (east vs. west).

The result of the election was a landslide victory for Cornelis-Christiandy, who this time won with more than 50% of the votes. Non-Muslims' support for this pair was universal and consistent irrespectively of the region in the province (compare Figure 5.14). The support for the three Malay candidates was much more geographically diversified; regions in West Kalimantan voted for their local Malay candidates. Tambul Husin-Barnabas Simin won in their home *kabupaten* of Kapuas Hulu (43%), but performed very poorly everywhere else (8% on average). Critically, however, the support for Tambul in Kapuas Hulu came only from the Muslim voters (compare Figure 5.15); the Christian population in the regency

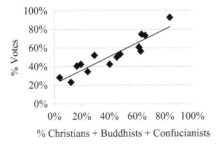

Figure 5.14 Support for Cornelis-Christiandy in the 2012 West Kalimantan gubernatorial election by regency and proportions of Christians, Buddhists and Confucianists combined

Source: Author's own compilation according to Badan Pusat Statistik (2010), *Pontianak Post* (2012a, 2012d).

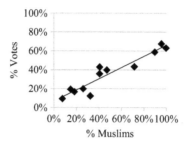

Figure 5.15 Support for Tambul-Barnabas in the 2012 West Kalimantan gubernatorial election in Kapuas Hulu and proportions of Muslims in sub-districts

Source: Author's own compilation according to Kementerian Agama (2010) and Constitutional Court Decision 70/PHPU.D-X/2012.

Note: At the time of writing, the election results tabulated by sub-district were available only from the Decision of the Constitutional Court. The decision, however, for the Kapuas Hulu regency included only results for about half of all sub-districts and only these are included in this analysis. There is no indication that the sub-districts not included in the decision text would show any different trends. Results correlated against Protestants and Catholics showed the same trend as the combined Christian vote.

supported Cornelis-Christiandy. Mobilization of the Kapuas Raya region failed, as no increased popularity of Tambul Husin (except his home regency) or Morkes Effendi within Kapuas Raya region could be appreciated.

Morkes Effendi-Burhanuddin won in Sambas and Kayong Utara and this ticket was the most popular Malay/Muslim candidates in general (26%); Armyn-Fathan won in Pontianak city (42%), but in total obtained only 15% of the votes. Cornelis-Christiandy won in Kubu Raya, Ketapang and Pontianak regency, despite these regencies' Muslim/Malay majority, just as they did in all non-Malay-majority regencies and Singkawang city.

While in the first direct gubernatorial election in 2007 virtually all Muslims voted for one of the Muslim candidates, in 2012 about 20% of Muslims voted for Cornelis-Christiandy, and they were evenly distributed across all regencies of West Kalimantan (see Figure 5.16). This is consistent with the general tendency of incumbents faring better in their run for the second term. In the 2012 election, we also observed that next to the religious split, the regional divisions among Malays were important to a greater extent than in 2007, proving that Malays retain at least two activated identities in their repertoires. However, the regional loyalties did not trump the religious identity.

5.5 Regent and mayoral elections

The presidential and gubernatorial elections as well as party preference in legislative elections showed that in these instances the Christian/Dayak–Muslim/Malay division remained the pervasive cleavage. In the next paragraphs, I will turn to regent and mayoral elections to test for activation of ethnic categories on the third tier of the Indonesian administration. Note that some potentially activated categories, chiefly Madurese and Javanese, cannot be appreciated given the data available and although in some regencies/cities (Kubu Raya, Pontianak) I can show attempts to mobilize the Madurese vote, I cannot show any evidence of how the Madurese actually voted. Similarly, Dayak linguistic and regional divides can be ascertained only by rough estimates based on a map of Dayak languages, and although in some cases the correlation is clear enough, in other cases no conclusions could be drawn because of lacking data. Nevertheless, a very careful study of each regency/city executive election shows several patterns of ethnic behaviour and unveils two ethnic dimensions activated at this level of politics which are not activated on other levels/in other elections.

To fully appreciate this point, one has to be aware of the very high, quite rare in social sciences, consistency of voters' behaviour in regencies in gubernatorial elections. If we take the 2007 election as a baseline (the 2012 one being distorted by strong incumbency effect), we see that in virtually every regency Cornelis' result was almost perfectly correlated with Christian faith of voters, and the trend is visible in regencies with as few as seven sub-districts (i.e. data points), of which Sekadau is a good example (Figure 5.16), but is replicated also in sizeable regencies with more than 20 sub-districts, like Kapuas Hulu (Figure 5.17). In

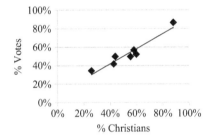

Figure 5.16 Support for Cornelis-Christiandy in Sekadau in the 2007 West Kalimantan gubernatorial election according to Christians' proportions in sub-districts

Source: Author's own compilation according to Kementerian Agama (2010) and Komisi Pemilihan Umum Provinsi Kalimantan Barat (2007).

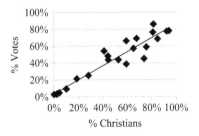

Figure 5.17 Support for Cornelis-Christiandy in the 2007 governor election according to Christians' proportions in sub-districts in Kapuas Hulu

Source: Author's own compilation according to Kementerian Agama (2010) and Komisi Pemilihan Umum Provinsi Kalimantan Barat (2007).

fact, Cornelis-Christiandy's support should be seen as coming from non-Muslims. The Singkawang results attest to that: the winning pair received votes proportional to numbers of combined Christians, Buddhists and Confucianists in sub-districts, indicating that the Chinese population also consistently supported the Dayak-Chinese pair.

Given, therefore, that for every regency the activated category dimension in gubernatorial election was religion (Muslim vs. non-Muslim), we ask whether the same dimension and set of categories is activated in the regencies when they elect their executive heads on the regency level. I will present several phenomena that can be observed in regency-level executive elections, which follow different patterns from the gubernatorial election: (1) in some *kabupaten* elections were devoid of the Muslim–Christian divide, demonstrating change in ethnic preferences of voters between elections on different tiers; (2) Catholic and Protestant categories are activated in those regencies that are Christian majority and have substantial numbers of Protestants (who are a minority compared to Catholics); (3) in some cases local linguistic/regional Dayak categories are activated in regency elections; a corresponding and likely Malay–Madurese division is to be expected and although evidence is weak, these cases are discussed accordingly; (4) almost all elections had most candidate pairs representing two main categories in the regency, or in other words, the candidates were paired to constitute an ethnic maximum-winning coalition; (5) an indirect consequence of (4) was the fact that it was often the vice-*bupati* who was able to win against his boss in the subsequent election, resulting in power changing hands between the two main categories; (6) a general trend, spanning through all regencies, is that incumbents win, often with increased margins of votes. In the following I discuss examples of each of these phenomena.

(1) Non-ethnic vote

The furthest east regency, Kapuas Hulu, has a substantial Malay plurality (49%). Muslims, however, are about 60% of the population, which indicates some Dayaks (next to the 3% Javanese) are also Muslims. Christians are mostly Dayaks. Kapuas Hulu regent elections have been dominated by Abang Tambul Husin, a Muslim Dayak/Malay, since 2000, when he won the indirect election in the legislature paired with a Christian Dayak. He was re-elected five years later in the first direct election. Paired with Yoseph Alexander (Christian Dayak) in 2005, Tambul competed chiefly against a fellow Muslim Baiduri Ahmad (paired with Antonius L. Ain Pamero, Christian Dayak). Tambul obtained 51% of the votes, Baiduri 38%. According to the ecological inference analysis run on these election results, Abang Tambul Husin was equally popular among Christians and Muslims (compare Figure 5.18). Baiduri-Pamero's support was also independent from religion, and the only pair with a Christian Dayak as regent, M Kebing L (paired with Kamsidi, a Malay), obtained a mere 6% of the votes.

In 2010 Abang Tambul Husin was not eligible for re-election, as he had served two full *bupati* terms. Abang Mahmud (AM) Nasir, son of Abang Tambul Husin, turned out to be the main contender in the election; he stood with Agus

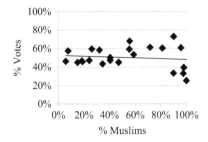

Figure 5.18 Support for Abang Tambul Husin–Yoseph Alexander by sub-districts and proportions of Muslims in the 2005 Kapuas Hulu regent election

Source: Author's own compilation according to Kementerian Agama (2010) and Komisi Pemilihan
Umum Kabupaten Kapuas Hulu (2005).

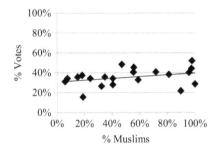

Figure 5.19 Support for Abang Mahmud (AM) Nasir–Agus Mulyana by sub-districts and
proportions of Muslims in the 2010 Kapuas Hulu regent election

Source: Author's own compilation according to Kementerian Agama (2010) and Komisi Pemilihan
Umum Kabupaten Kapuas Hulu (2010).

Note: Two outliers were excluded from the analysis: Bunut Hilir and Batang Lupar. These are home
sub-districts of AM Nasir and Agus Mulyana respectively, and the pair received 69% of the votes in
both, a much higher number than their average result in the regency.

Mulyana, a Christian Dayak, and won in the first round with 37% of the votes.
The Baiduri-Pamero pair stood again and were again the runners-up, with 30%.
Yoseph Alexander (incumbent vice-*bupati*, a Dayak, paired with Chairul Saleh, a
Malay) also contested, and received 20% of the votes.

This time around again the two main contenders were Muslim-Christian pairs
(or Malay-Dayak). The best Dayak-Malay pair in contest garnered a much more
significant portion of the vote than its counterpart in the previous election, but its
support was still lower than the number of Christians in the regency. AM Nasir
repeated his father's success to appeal to all religious groups equally. His support
grew only minimally in sub-districts with higher proportions of Muslims (compare Figure 5.19).

Therefore, Kapuas Hulu *bupati* elections follow their own trajectory, different from Kapuas Hulu voters' preferences in gubernatorial elections. While in gubernatorial polls the Christian vote in Kapuas Hulu was strictly behind the Christian candidate, in *bupati* elections significant numbers of voters opted for a Muslim candidate. As regent candidate, Abang Tambul Husin was able to attract voters from across all religious categories, although in the 2012 gubernatorial election, Tambul was the candidate of choice of *only* the Muslim electorate in the regency (Figure 5.15).

The 2008 Sanggau regency is a testimony to a similar pattern. Sanggau is a Dayak-majority *kabupaten* with 62% of Dayaks; Malays are 20%, Javanese 9% and Chinese 3%. Muslims account for about 33%, and Buddhists and Confucianists for about 3%. Catholics comprise 50% of the entire population, while Protestants are 16%. The 2008 *bupati* election in Sanggau had two rounds; ironically, just before the election the electoral law was changed to increase the winning threshold for the first round to 30%; had the law remained unchanged, there would have been only one round.

Yansen Akun Effendy became *bupati* of Sanggau in 2003 in the last indirect election. He was elected, as Hui established, as a Dayak leader, thanks to Dayak votes in the regency legislative assembly (DPRD) (2011, 294). However, because of his partly Chinese ancestry,[18] "the Chinese of West Kalimantan have since recognized him as the first Chinese to become a *bupati* in Indonesia" (Hui 2011, 294). The 2008 election saw six pairs of candidates; four Dayaks (of whom two paired with Malays, two with other Dayaks), and one Malay (Setiman Sudin, paired with a Dayak), and a Sino-Dayak, Yansen Akun Effendy, the incumbent *bupati*, paired with a Malay.

Yansen's support in the election was not significantly dependent on followers of any religion and he fared well in all sub-districts (Figure 5.20). Arguably thanks to this consistent and non-religious support, he won the first round, obtaining 28% of the votes. The runner-up of the first round, Setiman Sudin paired with Paolus Hadi (a Christian), attracted mostly Muslim votes (see Figure 5.21). The election went into the second round.

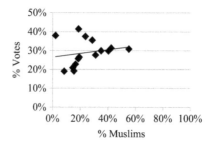

Figure 5.20 Support for Yansen-Abdullah according to proportions of Muslims in sub-districts in the first round of the 2008 Sanggau regent election

Source: Author's own compilation according to Kementerian Agama (2010) and Komisi Pemilihan Umum Kabupaten Sanggau (2008a).

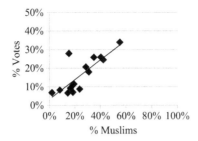

Figure 5.21 Support for Setiman Sudin–Paolus Hadi according to proportions of Muslims in sub-districts in the first round of the 2008 Sanggau regent election

Source: Author's own compilation according to Kementerian Agama (2010) and Komisi Pemilihan Umum Kabupaten Sanggau (2008a).

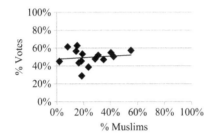

Figure 5.22 Support for Setiman Sudin–Paolus Hadi in the second round of the 2008 Sanggau regent election according to proportions of Muslims in sub-districts

Source: Author's own compilation according to Kementerian Agama (2010) and Komisi Pemilihan Umum Kabupaten Sanggau (2008b).

Yansen Akun Effendy's total number of votes went from 61,282 in the first round to 104,899 in the run-off. However, Setiman Sudin's increase in support was even more remarkable; from 43,094 votes in the first round, he increased to 109,942. Significantly, the second round of this election was not cleaved along religious lines. Setiman's strong Muslim support in the first round was entirely absent in the second round. In fact, analyses of both teams' performance in sub-districts revealed no religious pattern (see Figure 5.22). The Sanggau election shows, therefore, that one candidate's constituency can change significantly between two rounds of the same election, and switch from a strictly ethnic vote to entirely non-ethnic.

In the Sintang regency we could observe a similar switch as in the case of support for Setiman Sudin, but between two consecutive elections. Sintang is a Dayak- and Christian-majority regency. Muslims are 37%, while Catholics and Protestants are 38% and 24%, respectively. In the first direct regent election, the winning pair, Milton Crosby and Jarot Winarno (a Protestant Dayak and a Javanese), fared equally well among the religious categories (Figure 5.23).

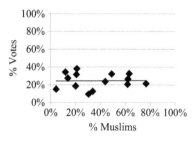

Figure 5.23 Support for Milton Crosby–Jarot Winarno in the 2006 Sintang regent election
according to proportions of Muslims in sub-districts

Source: Author's own compilation according to Kementerian Agama (2010) and Komisi Pemilihan
Umum Kabupaten Sintang (2006).

The 2010 Sintang election showed how, if a ruling regent–vice-regent pair
splits, one of them may turn to represent one ethnic category, and the other may
appeal to another, although earlier – together – they were able to attract the full
spectrum of voters. During the 2010 election, Milton refrained from brandishing
his ethnic affiliation (in the ballot paper photo he wore a Western-style jacket and
no ethnic markers), while his Dayak running mate displayed Dayak headgear.
Milton's win was chiefly due to Christian (and visibly stronger Protestant) and
Dayak votes (Figure 5.24). Jarot Winarno, incumbent vice-*bupati* and Javanese
paired with Kartiyus, a Malay, appealed visibly to Muslims, as this ticket's support came chiefly from this religious category (compare Figure 5.25).

The Kapuas Hulu, Sanggau and Sintang cases display the change from ethnic
appeal to non-ethnic appeal (or the other way around) between either elections on
different tiers, or two rounds of the same election or subsequent elections on the
same tier. The Sintang election also shows how candidates alter their ethnic appeal
depending on the strategy of the main contender. Milton Crosby (a Christian) ran
for the first election paired with a Muslim to appeal to the full religious spectrum,
but chose another Dayak (albeit a Catholic, creating a Protestant–Catholic team)
when challenged by a Muslim-only pair of strong contenders.

(2) Protestant–Catholic division

The Protestant and Catholic identities were activated in several sub-districts. In
the first round of the 2008 Sanggau election, Catholics had stronger preference
towards Andeng Suseno with Daniel Kwetono Djiono (Figure 5.26) and Thadeus
Yus with Petrus David (Figure 5.27), while Protestants were indifferent towards
these candidates. Protestants, on the other hand, voted for Krisantus Kurniawan,
towards whom Catholics were indifferent.

The Sintang elections are also an example of activated Protestant and Catholic categories. Figure 5.24 shows that Milton Crosby's popularity was clearly

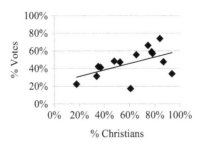

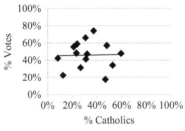

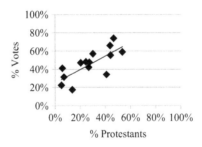

Figure 5.24 Support for Crosby-Ignatius according to religious proportions in sub-districts in the 2010 Sintang regent election

Source: Author's own compilation according to Kementerian Agama (2010) and Komisi Pemilihan Umum Kabupaten Sintang (2010).

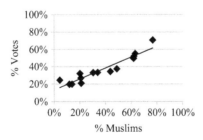

Figure 5.25 Support for Winarno-Kartiyus according to proportions of Muslims in sub-districts in the 2010 Sintang election

Source: Author's own compilation according to Kementerian Agama (2010) and Komisi Pemilihan Umum Kabupaten Sintang (2010).

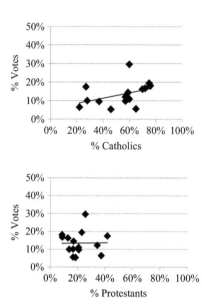

Figure 5.26 Support for Andeng Suseno–Daniel Kwetono Djiono according to the religious composition of sub-districts in the first round of the 2008 Sanggau regent election

Source: Author's own compilation according to Kementerian Agama (2010) and Komisi Pemilihan Umum Kabupaten Sanggau (2008a).

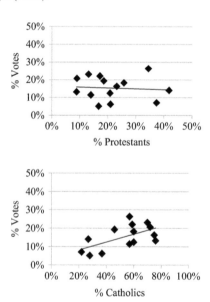

Figure 5.27 Support for Thadeus Yus–Petrus David according to the religious composition of sub-districts in the first round of the 2008 Sanggau regent election

Source: Author's own compilation according to Kementerian Agama (2010) and Komisi Pemilihan Umum Kabupaten Sanggau (2008a).

correlated to numbers of Protestants in districts (Catholics were indifferent towards him). Also in Sintang, Elyakim Simon Djalil's support grew along with numbers of Protestants (Figure 5.28). Mikail Abeng–Muhammad Yusuf's support was positively correlated with Catholics and negatively with Protestants (Figure 5.29).

(3) Linguistic division among Dayaks; Malay–Madurese division

A further important facet of the Sintang elections is the fact that the Catholic–Protestant division is almost co-terminus with geographical and linguistic divisions in the regency, and I traced similar patterns in Melawi, discussed later. A map provided by the Institut Dayakologi showing the distribution of Dayak sub-categories was used to establish the distribution of Dayak linguistic categories in the *kecamatan* (sub-districts). Although the method is far from precise, the trend is clearly visible. In the first round of the 2006 *bupati* election, the *kecamatan* with Ketungau as majority (Ketungau Hilir, Ketungau Hulu and Ketungau Tengah) voted for the fellow Ketungau and Protestant Elyakim Simon Djalil (Figure 5.28). Another candidate, Mikail Abeng with Muhammed Yusuf, was supported mostly by the Uud Danum Dayak category in the Serawai and Ambalau sub-districts;

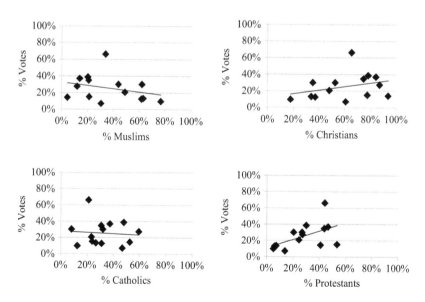

Figure 5.28 Electoral support for Elyakim Simon Djalil–Ade Kartawijaya in the 2006 Sintang regent election according to religious proportions in sub-districts

Source: Author's own compilation according to Kementerian Agama (2010) and Komisi Pemilihan Umum Kabupaten Sintang (2006).

Note: The outlier does not change the general trend in any significant way. However, note that it corresponds to Djalil's home area, Ketungau Hulu.

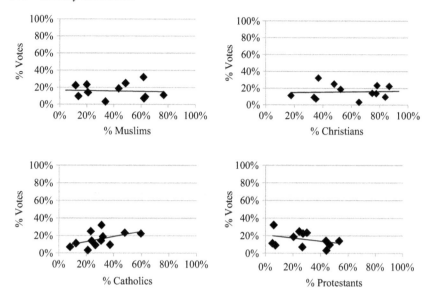

Figure 5.29 Support for Mikail Abeng–Muhammad Yusuf in the 2006 Sintang regent election according to religious proportions in sub-districts

Note: Two sub-districts, Serawai and Ambalau, Mikail Abeng's region of origin, were excluded from this analysis as outliers; the pair obtained close to 60% of the votes there. These votes, I argue, were for Mikail Abeng as a Uud Danum, not for Mikail Abeng as "Catholic", "Christian, "non-Muslim" or "non-Protestant". If the two sub-districts were included, Muslims would show a strong but inconsistent negative correlation, Christians positive, Catholics stronger positive and Protestants stronger negative.

Source: Author's own compilation according to Kementerian Agama (2010) and Komisi Pemilihan Umum Kabupaten Sintang (2006).

additionally his support grew along with Catholics' numbers in sub-districts, and was slightly negatively correlated with Protestants (Figure 5.29).

The Catholic–Protestant, combined with geographical, division was less visible in the second *bupati* election. By January 2010, Elyakim Simon Djalil devoted his efforts to creating a new *kabupaten*, Ketungau, out of the three northern-most sub-districts of Sintang (*Kalimantan-news* 2012). Not coincidentally, the population of these three sub-districts had lent their support to Djalil in the 2006 *bupati* election. Presumably because of the prospect of creating the new regency, Elyakim Simon Djalil refrained from running in the Sintang *bupati* election.[19] Mikael Abeng this time stood for election with a new running mate, Suryanto Tan-jung, the head of the Ikatan Masyarakat Kayan (Association of the Kayan People) (*Kalimantan-news* 2011). Kayan is a Dayak category constituting significant num-bers of the population in the Kayan Hilir and Kayan Hulu sub-districts. Hence, the Abeng-Tanjung candidate pair should be seen as an Uud Danum–Kayan Dayak coalition which coincides geographically with the four easternmost sub-districts of Sintang. Significantly, this pair's result in the four *kecamatan* averaged 30%, while in the remaining sub-districts it was only 7%.

Landak and Sekadau are two regencies in which geographical/linguistic divisions among Dayaks could be appreciated. In these regencies Catholics are the majority among Dayaks and the Catholic–Protestant cleavage was absent here. The vote was split geographically in a way that corresponds to linguistic divisions among Dayak sub-categories. The Sekadau regency's population is 58% Dayaks, 26% Malays, 9% Javanese and 3% Chinese. Muslims constitute 38% of the population in Sekadau, Christians 61%, and about 1% are followers of Buddhism and Confucianism (Badan Pusat Statistik 2010). Forty-seven percent of the population is Catholic and 14% Protestant. Among the Dayaks, 25% are Dayak Mualang, who are concentrated in the north of the *kabupaten*. Belitang Hilir and Belitang Hulu are two sub-districts in which the Mualang are predominant; there are also significant numbers of Mualang in Belitang. Among other Dayak categories, Dayak Ketungau are about 15% and are mostly concentrated in the Belitang sub-district, squeezed between Belitang Hilir, Belitang Hulu sub-districts and the Sintang regency. Another significant Dayak category (10%) is the Mentuka, concentrated in the Nanga Taman and Nanga Mahap sub-districts in the south.

Both *bupati* elections (in 2005 and 2010) in Sekadau were won by Simon Petrus, a Dayak and Christian. In 2005, paired with Abun Ediyanto (of mixed Chinese and Dayak parentage), Simon gathered 28% of the votes in total, and won in three out of seven sub-districts (Belitang, Belitang Hilir and Belitang Hulu, all of them in the north of Sekadau). His performance in this election was strongest in the Belitang Hulu and Hilir sub-districts, as well as in Belitang, or, in other terms, in the north of the regency. In three other *kecamatan*, Simon Petrus came third and in one sub-district, he came second. The runner-up, Stefanus Masiun (with Petrus Langsung, both Dayaks), received 22% of the total votes, winning in two southern sub-districts (Nanga Mahap and Sekadau Hulu). A Malay, Benny Pensong (running with a Christian, Hugo Agato) won in one sub-district (Sekadau Hilir), and received 20% of the votes; another Muslim candidate received 15% of the votes. Aloysius Aleksander (paired with Norbertus, both Dayak Christians) obtained 16% of the votes and won in the Nanga Taman sub-district. The combined vote cast for the two Muslim candidates amounted to 34.7%, which is only slightly less than the entire Muslim population of Sekadau, and the support for the Muslim candidates grew along higher numbers of Muslims in sub-districts (Figure 5.30). Otherwise, ecological inference analysis of the vote against ethnic numbers brought weak results, which would be consistent with a cleavage running along the geographical north–south (or Dayak sub-category) line.

Therefore, in the 2005 *bupati* election in Sekadau there were two cleavages: north versus south and Christians versus Muslims. Although the precise sub-category background of the other Dayak categories is unknown (and the distribution of these smaller, fragmented southern categories is difficult to track), it is clear that the vote in the south was split between Aloysius Aleksander and Stefanus Masiun, while the north was strongly behind Simon Petrus.

In the 2010 election Simon Petrus (this time paired with Rupinus, a Catholic Dayak) won in a landslide. He won in all sub-districts with very high margins and obtained 56% of the votes. This time his strongest contender was the incumbent

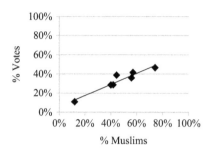

Figure 5.30 Combined support for the two Muslim candidates in the 2005 Sekadau regent election according to proportions of Muslims in sub-districts

Source: Author's own compilation according to Kementerian Agama (2010) and Komisi Pemilihan Umum Kabupaten Sekadau (2005).

vice-*bupati*, Abun Ediyanto. The closest Abun came to challenging Petrus was in Sekadau Hilir, where Abun obtained 27% of the votes, against Petrus' 44%. Stefanus Masiun and Pensong both ran again in 2010, both paired with new running mates. Their performance was visibly poorer than in the first election and both came third and fourth, respectively (after Simon Petrus and Abun Ediyanto),[20] in all sub-districts. None of the candidates' support was strongly dependent on any particular religious category. The north–south divide was still detectable in the 2010 electoral results. The two southern *kabupaten* (Nanga Mahap and Nanga Taman) were the only two where Simon Petrus obtained less than 50% of the votes.

The Landak regency shows a similar discrepancy in candidates' popularity depending on the Dayak sub-category. The first direct *bupati* election in Landak took place in 2006. Cornelis, of Kenayatn Dayaks origin, the biggest and currently most influential Dayak sub-category in the province, was contesting on a ticket with Adrianus Asia Sidot; the incumbent vice-*bupati*, Nehen, was running with Yohannes Bahari. Syahdan Anggoi (with Chritianto Syam) and Kartius (with Yohanes Meter) were the other two contesting tickets. Cornelis-Adrianus obtained 46% of the total votes; the runner-up, Syahdan Anggoi, received 25% of the votes; Kartius came third with 15%, while Nehen was last with 14%.

The geographic distribution of the support was significant. Nehen won in two sub-districts: Air Besar and Kuala Behe, which are areas of non-Kenayatn Dayaks; Cornelis won with enormous margins in those sub-districts which are Kenayatn-majority: Sengah Temila, Mandor, Menjalin and Mempawah Hulu. The election was visibly cleaved along the Kenayatn–non-Kenayatn line. In 2007, Cornelis ran in the gubernatorial election. His result in Landak, expectedly, was the highest among all regencies and only in three sub-districts in Landak was Cornelis' result lower than 80%: in Sebangki, which is a Muslim-majority *kecamatan*, as well as in Air Besar and Kuala Behe, which are non-Kenayatn sub-districts.

Attempts at capitalizing on the Madurese identity in politics have been observed in those regencies/cities in which Madurese constitute significant numbers: Kubu Raya and Pontianak regencies, as well as Singkawang and Pontianak cities. As pointed out earlier, no data showing numbers of Madurese, Javanese or Malays by sub-district is available and there is no evidence of how the Madurese candidates resonated among the voters. However, in the first direct mayoral election in the city of Pontianak, the winning pair was the sole Malay-Madurese pair (other candidates were paired between Malays and Chinese, Chinese being the other significant category in the city, or were Malay-only candidates). In the next election, the Madurese vice-mayor ran as a mayoral candidate with a Chinese deputy, but lost to the incumbent mayor, who was paired with another Malay.

In the regency of Pontianak, the main battle in the first direct election was between tickets Ria Norsan–Rubijanto and Agus Salim–Muhammad Saleh. The first pair won in the first round with more than 47% of the votes, while the latter received 34%. One of the campaign issues were flyers which claimed that Ria Norsan was not only born in Sambas, but also that he profited financially from the 1999 anti-Madurese riots (*Pontianak Post* 2008). These facts could have been detrimental to his image among the Madurese community. Ria Norsan denied all the information, although he was in fact born in Singkawang, which at the time of his birth in 1967 was part of the *kabupaten* Sambas. In order to emphasize his commitment to the Madurese electorate, Ria Norsan and his deputy, Rubijanto, vowed in front of an imam hailing from Madura that, if elected, "they would carry out development without differentiating [between ethnic categories]" (*Pontianak Post* 2008). Ria Norsan–Rubijanto won the election and Ria Norsan was re-elected (with a different Malay deputy) in 2013. Also in the 2012 Singkawang election a Madurese vice-mayor candidate may have been the game changer. Although the 2007 election was a plain Chinese versus Malay affair, in the 2012 election not only was there only one Malay mayoral candidate, which prevented a Malay vote split, but he paired with a Madurese, which likely secured him votes from this minority.

In the first *bupati* election in Kubu Raya, a Madurese candidate, Suhri Maksudi (paired with a Malay, Lendeng Syahrani), oriented his campaign on mobilizing the Madurese. He received official endorsement from important figures of the East Javanese and Bugis communities, as well as from people related to *pesantren* (Islamic boarding schools) and *ulamas* (*Pontianak Post* 2010). However, this strategy brought poor rewards, as Suhri-Lendeng obtained only 4% of the votes.

Despite the weak or non-existent evidence of actual support for these candidates, I showed that the Madurese category is being activated in elections in those areas where Madurese represent significant numbers among the voters. It is likely that a similar process is happening to the Javanese category (I discuss such a case in the Ketapang regency later), but as the Javanese are much more dispersed in West Kalimantan and their absolute numbers are even lower than the Madurese, little can be found to support any claim of Javanese category activation.

(4) Candidate pairs as maximum-winning coalitions

Most candidates in regencies across West Kalimantan were maximum-winning coalitions and in only a few cases candidates opted for an exclusive coalition, although the strategy proved successful in both gubernatorial elections. In regencies where Christian and Muslims were in substantial numbers (Kapuas Hulu, Sintang first election, Melawi, Ketapang), most pairs include a Christian and a Muslim and the advantage of one category over another is displayed by who is the main candidate and who is the deputy. In Singkawang the main ethnic split is between the Chinese and Malays and most candidate pairs included representatives of both ethnic categories; in the first direct election in Singkawang there was also a Malay-Dayak pair, although the numbers of Dayaks in Singkawang are negligible. In the city of Pontianak, where Malays, Chinese and Madurese are all substantial with Malays being a majority, candidate pairs included all three categories in different combinations, including a Madurese-Chinese pair in the second election. In Kubu Raya, a regency with high numbers of migrants from other regions of Indonesia, the winning coalition in the first election was a Malay with a Javanese Catholic (!) – an ethnic coalition that could attract several categories otherwise difficult to be represented by merely two individuals. In the second election in 2013, a Bugis (Muslim) candidate paired with a Dayak (Christian) won the Kubu Raya *bupati* seat; not only did this ticket beat the sitting *bupati* (the only such instance in West Kalimantan in the observed period), but also did so in the first round.

Naturally, this pattern was hardly observed in regencies with very high majorities of one category. In Sambas, a regency with more than 90% of Malays, all but one candidate pair in two elections included only Malays; in Landak, a regency that has more than 90% of Dayaks of whom a significant majority is Catholic, all candidate pairs were Dayak and Catholic.

There is, however, the logical temptation to break the pattern and create an exclusive coalition. This was the case of Sintang, where the second election was a fierce competition between two exclusive (Christian vs. Muslim) candidate pairs, although the first election was won by a Christian and Muslim pair. In the city of Pontianak, in the first election it was a Malay-Madurese coalition that garnered the most votes, but the winning pair broke up before the second term and the Malay incumbent ran along with another Malay, while the incumbent vice-mayor and Madurese ran with a Chinese mate. The Malay team won. Also in Sanggau the first election was won by a Muslim-Christian pair, but the second was won by a fully Christian pair. In Sekadau, Muslims have not been represented on the winning ballot in either election, despite constituting almost 38% of the population. In Singkawang, although in the first election all candidate pairs were maximum-winning coalitions (Chinese-Malay or Malay-Chinese and one Malay-Dayak), in the second election none was (Chinese paired with Chinese, while the winning pair was Malay with a Madurese).

Two points are due; firstly, voter support is in almost all cases correlated to the ethnic identity of the regent/mayor candidate, and there is rarely any sign that voters are drawn to the deputy's identity. Exceptions are well represented by a couple of Dayak-majority sub-districts in the overwhelmingly Malay Sambas regency. These two sub-districts voted for the only Malay-Dayak candidate pair in the running in 2011, and this must be attributed to the Dayak identity of the deputy candidate on the ballot. The pair's popularity was abysmal in all other, purely Malay, sub-districts in Sambas. In most other cases, the running mate's identity has merely token value and points to an established tradition, discussed in earlier chapters, of seeking power-sharing solutions.

Secondly, it is easy to establish that exclusive, minimum-winning coalitions fare better in elections; the gubernatorial election, the second Pontianak mayoral election or the second Sintang election point in this direction (although in the latter two cases the winners were also incumbents). As was mentioned, West Kalimantan displays a tendency to power-sharing schemes, which may be a self-reinforcing phenomenon through two parallel channels. One is the politicians' expectations that this is what voters prefer and feel safe about. Ethnic strife, also violent, is not foreign to the region and voters likely perceive the multiethnic running teams as a preventive measure and a symbol of peaceful coexistence. Another reason that the maximum-winning coalitions may be reinforcing themselves is the fact that if the winning coalition includes both main categories, the spoils of power will be shared, which is something politicians may prefer, regardless of voters' attitude. In the following I will argue that these coalitions likely lead to arresting activated categories around two identities on one dimension over a period of time (e.g. Christian-Muslim or Chinese-Malay)

(5) Ethnic categories alternate as executive heads

As a consequence of the phenomenon of multiethnic candidate pairs, one has to note the fact that in some cases power was alternated between the two main categories from one electoral cycle to another. The 2002–2007 Singkawang mayor and vice-mayor were Awang Ishack (Malay) and Raymondus Sailan (Dayak). In the 2007 mayoral election in Singkawang, there were five candidate pairs, of which in four the mayoral candidate was a Muslim; the vice-mayor candidates were Chinese in three cases and Dayak in one pair. The fifth pair had a Chinese as a mayoral candidate and a Malay as a vice-mayoral candidate. Because the Malay vote was split between the three candidates, Hasan Karman, the only Chinese candidate, won the 2007 election. The 2012 election was in many ways the opposite: among the four candidate pairs, in three a Chinese (all paired with another Chinese) was the mayor candidate, and only in one a Malay (the 2002–2007 mayor, paired with a Madurese). This time around, the Chinese vote was split between the Chinese candidates, while the Muslim vote was united around the Malay-Madurese team. Singkawang is therefore a city that, because of its ethno-religious split, fixes the voters' preference around the religious differences (Muslim vs. non-Muslim). In Singkawang the 2012 mayoral election took place on the same day as the gubernatorial election. While the Malays voted as a united front for a Malay as mayor (who won in the end), in the gubernatorial election the Malay vote was split between three candidates, which allowed Cornelis-Christiandy to capture the highest number of votes in the city.

Therefore, Singkawang suggests that in cities/regencies with almost even split between two categories, power will likely switch hands between these two categories in subsequent election. Ketapang is another example of this trend, although so far the evidence is based on only three electoral cycles. Ketapang is a regency with a 70% Muslim majority. Among the Christians, Catholics are about 70%. Interestingly, 49% of Ketapang residents also declared themselves Dayaks. There are 31% Malays, 10% Javanese and 5% Madurese, which leaves about 20% of the population to be Muslim Dayaks.

The first direct election was won by Morkes Effendi, incumbent *bupati* 2000–2005. The then vice-*bupati*, Lourentius Majun (Dayak), ran as a *bupati* candidate himself on a ticket with a Malay (Abul Ainen). The third pair was a Malay (Gusti Sofyan Afsier) with a Dayak (Paolus Lukas Denggol). Effendi was re-elected (43% of the votes) with a new running mate, Henrikus (a Christian Dayak). Lourentius came second (with 32%), and Gusti-Paolus obtained 26% of the votes. In terms of electoral cleavages, the election reflected the religious Muslims versus Christians split. Figure 5.31 shows the consistent correlation between the proportions of Christians and support for the sole Christian-led ticket in this election (Lourentius Majun–Abul Ainen).

In 2010, Henrikus, the sitting vice-*bupati*, a Catholic Dayak, and Boyman Harun (Malay) were the main contenders. Another candidate pair was Ismet Siswadi, a Javanese, who ran in the election as an independent on a ticket with Suhermansyah (Malay). The main contenders were, however, Yasyir Ansyari, the son of the incumbent *bupati* and Ketapang Golkar chair Morkes Effendi, and Martin Rantan (head of the Ketapang chapter of Dewan Adat Dayak). Martin Rantan during a rally said, "On the ballot paper I appear wearing traditional Dayak dress, while Yasyir Ansyari wears traditional Malay dress. Maybe as a pair we are a Malay and a Dayak, but if we are entrusted the confidence [and win the election], we won't be merely the *bupati* and vice-*bupati* of the Malays and Dayaks. To the contrary, [we will be] *bupati* and vice-*bupati* to all ethnicities, religions and groups [golongan]" (*Harian Equator* 2010b).

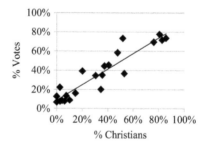

Figure 5.31 Support for Lourentius Majun–Abul Ainen in the 2005 Ketapang regent election according to proportions of Christians in sub-districts

Source: Author's own compilation according to Kementerian Agama (2010) and Komisi Pemilihan Umum Kabupaten Ketapang (2005).

Note: Each of the Muslim candidates was supported by Muslims and their electoral results put together gave mirror images of those presented in this figure. Catholics and Protestants showed the same trends as the two categories put together.

Contrary to this statement, however, during the campaign Yasyir's father, the incumbent *bupati* and the head of Majelis Adat Budaya Melayu (MABM) donated 20 million Indonesian rupiah (equivalent to about USD 2,000) to a mosque and on this occasion said, "Islam and Malayness are inseparable. The Malay tradition cannot stray away from the scriptures of Quran" (*Harian Equator* 2010a). Morkes Effendi and his son are from an aristocratic Malay family, which was reflected in Yasyir's ballot paper picture. The yellow shirt and headgear with golden embroidery left no doubt as to his aristocratic Malay background. Most interestingly, however, one of the most prominent faces of the campaign was Lourentius Majun (*Harian Equator* 2010d), vice-*bupati* under Morkes Effendi during the 2000–2005 term. Majun, after parting ways with Effendi in 2005, returned to Effendi's camp for the 2010 election and toured the *kabupaten* campaigning for the Yasyir-Martin team (*Harian Equator* 2010c).

There is little surprise in the fact that sub-districts with high proportions of non-Muslim voters supported Henrikus-Boyman (see Figure 5.32), and sub-districts with more Muslims voted for Yasyir-Mantan. The popularity of the Mecer-Jamhuri candidate pair increased with a higher proportion of Christians, while the popularity of Ismet-Syarmashyah correlated with increased proportion of Muslims (because of space constraints these graphs are not shown here).

The results of the first round were so close that Yasyir Ansyari and Martin Rantan, short by 0.13% of the winning 30%, challenged the result in the Constitutional Court. The court ruled that the result was valid and the second round had to be conducted. In the run-off Henrikus-Boyman won in what in these conditions should be deemed a landslide; these candidates gained most of the votes from the two pairs that lost in the first round and increased their absolute number of votes by almost 100% (from 58,656 votes in the first round to 116,079 in the second). Yasyir-Martin's gain was less than 50% (from 65,607 to 94,052). Although the vote was visibly cleaved along the religious line, Henrikus-Boyman obtained on average about 25% more votes than the numbers of Christians (Figure 5.33).

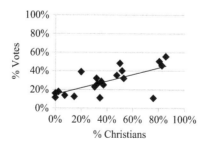

Figure 5.32 Proportion of Christians in sub-districts and support for Henrikus-Boyman in the 2010 Ketapang regent election (first round)

Source: Author's own compilation according to Kementerian Agama (2010) and Komisi Pemilihan Umum Kabupaten Ketapang (2010a).

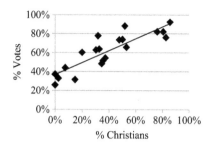

Figure 5.33 Support for Henrikus-Boyman in the run-off of the 2010 Ketapang regent election according to proportions of Christians in sub-districts

Source: Author's own compilation according to Kementerian Agama (2010) and Komisi Pemilihan Umum Kabupaten Ketapang (2010b).

A shift from one main ethnic category to another between two elections on the regent's/mayor's seat also took place in Melawi (a Muslim incumbent vice-*bupati* won the election after the Christian sitting *bupati* died), and in Sanggau (a Christian incumbent vice-regent ran for his boss's seat; the incumbent regent did not run for re-election), and these cases constitute less than half of the administrative units in West Kalimantan in which this pattern could be logically expected (one has to exclude these regencies with an overwhelming majority of one category: Sambas, Landak, Kayong Utara and Pontianak). This trend can be only observed with more electoral cycles.

(6) Incumbency effect

Although only two electoral cycles of direct elections in Indonesia are available for analysis at this time, some trends are already visible, especially if the analysis starts with the *bupatis* elected in indirect elections before the direct elections were introduced. The strong positive effect of incumbency could be appreciated in all but a few cases, and as incumbents some candidates tended to receive support from a wider ethnic constituency than in the first election. In other words, incumbency can trump ethnicity, as voters may prefer to choose the expected winner than their co-ethnic; as was the case in Landak and Sekadau, as well as in the gubernatorial election. As there is a limit of two terms for all executive offices in Indonesia, one should expect a vacuum after a popular officeholder ends the second term. In most cases on the regency level this problem is solved by the vice-regent stepping in as the main candidate. Vice-regents won in Ketapang, Landak, Sambas, Bengkayang, Melawi and Sanggau. In Kapuas Hulu the son of the sitting regent won the election after his father finished two terms, although he had to compete against the incumbent vice-regent. The opposite happened in Ketapang, where the *bupati*'s son's bid failed and the vice-*bupati* came out winning against the son of his former boss.

Few incumbents, however, failed to secure the second term. Muda Mahendrawan in Kubu Raya has been the only *bupati* elected without party support in the province (a rare occurrence across Indonesia in general), but after the first term he lost to a contender by a narrow margin. Hasan Karman in Singkawang and Setiman Sudin in Sanggau did not run for the second term; Setiman's deputy won the next election, while in the case of Singkawang, the office was secured by the mayor who had held the office during the term before Hasan Karman. In Kayong Utara and Pontianak regency, the incumbents won the second term with a lower margin of votes than in the first election.

5.6 Conclusions

Arguably, the Reformasi period in Indonesia represents a specific time in the country's history, when institutions are more fluid than in other times. This research looked at a very time-constrained set of events, and indeed the assumed ethnic identity change could only be observed in two broad dimensions: firstly, between the New Order and the liberal democracy period, and secondly, between consecutive elections since Suharto's stepdown. During the New Order and earlier, there were, I argued, three main categories activated in West Kalimantan: Malay, Dayak and Chinese. For a short time, a division between Catholics and Protestants among the Dayaks was activated (compare Dayak Unity Party's history presented earlier), and in the first Reformasi years, Madurese as a category was activated for the first time through violent events that involved this category. In the post-Suharto period, Catholics and Protestants were (again) activated as separate categories in direct regent elections, and there were attempts – also chiefly on the third tier of administration – to activate Madurese, but the response of the community to these efforts could not be ascertained.

Since the direct executive elections were introduced in 2004, elections on all levels were held twice. In four *kabupaten*, the second election results in ethnic terms were the reverse of the first: Ketapang, Melawi, Sanggau and Singkawang are almost evenly split between non-Muslims and Muslims, and in all four the second election was won by the category that lost in the first election. This phenomenon indicates that in the said regencies the Muslim versus non-Muslim cleavage was strengthened by the repeated competition along the same ethnic divide. In general, however, it should be expected that in those *kabupaten* where the main numerical split in the ethnic distribution is quite even between Muslims and non-Muslims, the *bupati* seat may forever be exchanged between the two religious categories. Especially in the pattern that is now being established (see Ketapang, Melawi), the *bupati* and vice-*bupati* are of a different religious background, and the vice-*bupati* challenges his boss after the first term. Alternatively, if the vice-*bupati* joined only for the second term, when his boss is not eligible for re-election, the vice-*bupati* wins just as an incumbent would. This pattern almost ensures that the bupati's position will change hands between Muslims and non-Muslims, endlessly repeating the activation of these religious categories.

A counterexample is the case of Kapuas Hulu. A Muslim *bupati* with a Christian deputy was replaced by another Muslim regent with a new Christian vice-regent.

The incumbent deputy had run for the *bupati* office, but lost to the son of the sitting *bupati*. This way, for three consecutive terms the regent was a Malay and his deputy a Christian – a counterexample to the pattern of changing regent offices between Malays and Christians after one or two terms. Although it is unlikely that every regent will have a suitable offspring to run for his office after the allowed two terms, there are precedents proving that nepotism may be interfering with what looks like patterns of power sharing. Note, however, that in 10 out of 14 third-tier units, the *bupati* and mayor pairs combine politicians of different ethnic backgrounds. In only four *kabupaten* (Sambas, Landak, Pontianak regency and Sekadau) the winning pairs were from the same ethnic categories. Except Sekadau, these *kabupaten* are also the most homogenous, i.e. the tickets, despite including only one category, were still maximum-winning coalitions. In Sekadau, the entirely Dayak and Catholic category was a minimum-winning coalition. In general, however, in most *bupati* and gubernatorial elections there was a strong preference for the tickets to be inclusive and maximum winning, i.e. combining two categories.

The Sintang *bupati* elections pointed to an interesting feature of the dual ticket. Support for the same regent candidate came in two consecutive elections from different ethnic quarters, visibly depending on the ethnic membership of the running mate. In the first election, the winning pair (Protestant Dayak with Muslim Javanese) was supported almost evenly by all religious categories. When in the subsequent election the incumbent pair split and the Protestant ran together with a Catholic, they won on the Christian vote, but overwhelmingly from the Protestant part of it. The Javanese incumbent vice-regent, now paired with a Malay, was a close runner-up, and was supported clearly by Muslims. Therefore, a change of the running mate and splitting of a maximum-winning coalition into two exclusive, minimum-winning ones does induce a shift of loyalties in voters. In general, however, the incumbent advantage was shown to have a neutralizing effect on ethnic loyalty (see the gubernatorial elections, the Landak and Sekadau *bupati* elections).

Can any more ethnic variety be expected in the subsequent elections in West Kalimantan? Or should we expect that categories activated during the first two cycles of executive elections will be perpetuated throughout the next terms? West Kalimantan is well worth further studies in order to test whether the current period of relatively dynamic changes is *just* a period (like it was in Sarawak in the 1960s and 1970s, when categories were newly activated and deactivated frequently), or whether it is a property induced by the particular institutional setting of Indonesia.

Assessment based on this research would suggest the first. It can already be seen that there is a certain amount of inertia in ethnic mobilization. Activated categories have the property of self-reinforcing, especially in the context of implicit mobilization. Implicit activation takes time, and the implicit message is less precise and not as well-targeted as an explicit message. Hence, once a party or a leader established its image as related to membership in a particular category, it will be difficult and time-consuming to shift to a different ethnic category as an appeal strategy. Note that the categories that have been identified most successfully over time and elections (Malay/Muslim, Dayak/Christian, Chinese) are also the most visible of all ethnic categories in West Kalimantan: they involve characteristic clothing, distinct personal names, well-established and prominent ethnic organizations and combine religious membership in them. Equipped with these elements, a candidate can unequivocally demonstrate whose ethnic interests he represents without having to say the name of the category, which arguably in the Indonesian context would be illegal.

The same property should be attributed to political parties' identification with particular categories. The ethnic outlook of parties (except for the explicitly religious parties) can be mostly established through implicit messages: candidates' ethnic membership shown through ethnic clothes, party leaders' membership in ethnic organizations, parties' consequent support for executive candidates of one particular category. Ethnic identity change induced through parties' mobilization strategy in this context cannot be too fast, as it takes time to convey the ethnic message through these implicit means. Although I claimed PDI-P and Golkar to be locally ethnic parties, obviously none of the parties' interviewed leaders would admit such ethnic commitment.

The 2012 election may have well been the last gubernatorial election of West Kalimantan in the province's current shape. Separation of the five eastern *kabupaten* will produce an entirely new set of majorities and minorities in the two units, and will emphasize other divisions in the new provinces that may have been dormant in the existing West Kalimantan. Therefore, on the provincial level it will not be possible to assess the presumed arrest of ethnic identity change due to repeated modes of category activation – in the next gubernatorial election we will likely be dealing with entirely new ethnic proportions in two provinces (West Kalimantan and Kapuas Raya), instead of one.

If the provincial split does not take place within the current five-year term of the governor, the next gubernatorial election of West Kalimantan would be of most interest for students of ethnic identity change. Cornelis and Christiandy will not be eligible for the next term and the existing Dayak-Chinese coalition may not survive the leadership void. The next West Kalimantan gubernatorial election – whether in its current shape or after division – is due in 2017. The current, extremely consistent Dayak support, irrespective of region, confession and Dayak linguistic sub-categories, for the Dayak Kenayatn Catholic governor is stunning. The absence of intra-Dayak strives must be seen as indication that non-activation of categories, or maintaining coherence of one category in the light of readily available dimensions along which the category could be internally split, is a powerful political tool. Simply put, both activation and non-activation of new categories are important and are phenomena worth watching in politics.

The multiplicity of elected offices in Indonesia was argued to offer an opportunity to induce multiple identities for each voter in West Kalimantan. This hypothesis was, however, only partly confirmed through the studies case. Presidential elections were shown to reflect party preference more than ethnic preference alone. As long as parties in West Kalimantan maintain their ethnic image, the presidential elections are likely to reflect the intra-province cleavage between

Muslims and non-Muslims, as main parties follow this religious split. Therefore, having taken the 2007 gubernatorial election as a baseline, we found no alternative categories activated in the presidential elections. The Muslim and non-Muslim vote was split in the presidential election in a very similar way to the split of the gubernatorial election.

The most universal cleavage in the regent elections was along the religious divide, but most regent elections saw the activation of additional categories that were not activated on other administrative levels. In several Dayak-majority regencies I discovered that Catholics and Protestants had differing political preferences. The Catholic and Protestant identities were not activated in the gubernatorial elections, although it may be a result of the particular set of candidates competing in the gubernatorial elections. With the help of a very crude method – by comparing distribution of Dayak linguistic categories represented on a map with the distribution of candidates' popularity – I was able to prove that in some *kabupaten* the vote was split between candidates from as many as three different linguistic categories within the regencies. The vote distribution coincided with region of origin (and Dayak sub-category) of the main contenders in these elections (Sintang, Sekadau and Melawi). In Melawi the geospatial difference in preferences coincided both with division between linguistic categories and Christian confessions.

Differing ethnic preferences expressed in different elections were most clearly visible in the case of Kapuas Hulu. The same candidate from the regency (Abang Tambul Husin, a Muslim Dayak/Malay) ran for two offices: regent and governor. Abang Tambul Husin won the regent election supported equally by Malays and Dayaks. He also performed strongly in the 2012 gubernatorial election in Kapuas Hulu, winning in the regency against the main contender. However, in this election Tambul's support came strictly from the Muslim quarters, while Dayaks voted for the Dayak-Chinese candidate pair. Interestingly, in the gubernatorial election Tambul Husin was also mobilizing on the regional dimension, calling for support from voters in the eastern regencies of the province. While the strategy failed, it shows that there are attempts to break the Muslim versus non-Muslim cleavage in West Kalimantan.

The dual ticket in executive elections was shown to produce two types of outcomes. One is maximum-winning coalitions: candidates of the two biggest categories are paired up and vie for votes from the entire ethnic spectrum of the constituency. So was the strategy of Malay-Dayak candidate pairs in the 2007 gubernatorial election as well as in Kapuas Hulu, Ketapang, Melawi regent elections. The other outcome is a minimum-winning coalition, made of one relatively big category (e.g. "Dayak" in both gubernatorial elections) and a smaller one ("Chinese" in the said elections); next to the governor winning candidates, also Pontianak city saw a similar case in its mayor election (Malay-Madurese winner pair), while first Sintang *bupati* election had a Javanese as the running mate of the winning candidate. The position of a king-maker is typical for minimum-winning coalitions; these numerically less powerful categories become of importance only in the case of an ethnically exclusive mobilization strategy. A maximum-winning coalition – of which an example are all Christian Dayak–Muslim Malay coalitions in West Kalimantan – eliminates the importance of those smaller categories, and emphasizes unifying and harmonizing properties of the dual ticket.

The relatively low winning threshold in executive elections (governor and regent), combined with the double ticket, clearly induces a search for minimum-winning coalitions and prevents the dichotomizing effect known from plurality elections. Multiple elections in West Kalimantan were shown to produce several (as many as eight!) candidate pairs, the pairs differing in their inclusiveness levels and the ethnic dimensions involved. The limited number of terms in executive positions has a similar effect; power is passed on to someone else, and even if it is the deputy who takes over from his boss, the enforced change offers an opportunity to reshuffle the ethnic combination that holds power.

The run-offs of *bupati* elections also produced important findings. In all three cases of the second round occurring in regencies in West Kalimantan, the second round's results reversed the order of the first round. Only one run-off, in Ketapang, led to what would be a commonsensical result – consolidation and dichotomization of the vote. In Sanggau, the comparison of candidates' performance in the two rounds showed that Yansen Akun Effendy's support remained independent from religious preferences in both rounds, while Setiman Sudin's support was strongly dependent on Muslims in the first round, and in the second round became non-communal. Therefore, elections with run-offs are another potential occasion of shift in ethnic identities. The small sample available for this analysis (only a small proportion of elections had a run-off) precludes drawing further conclusions.

Notes

1 PKB is officially a *Pancasila* party; however, being established and led by leaders close to Nahdlatul Ulama, the party's orientation towards Islam is difficult to hide (compare Woodward 2008, 42).
2 PDI did not automatically qualify for the 2004 election and could only contest after de-registration and registration under a new name. It did not obtain a seat in the 2004–2009 parliament.
3 Samson Darmawan, interview by the author, 1 March 2011; Karolin Margret Natasa, interview by the author, 5 March 2011; Agustinus Alibata, interview by the author, 14 March 2011.
4 Asya'ari, interview by the author, 18 March 2011; Adang Gunawan, interview by the author, 16 May 2011.
5 Muda Mahendrawan, interview by the author, 12 May 2011.
6 It was the 2013 Pontianak regency *bupati* election, when both parties backed the sitting *bupati*, Ria Norsan. The ticket enjoyed the support of PDI-P, Demokrat, Golkar, PAN and PPP, or all parties that mattered (*Tribun Pontianak* 2012b).
7 These results refer to the provincial DPRD election. There were only minor inconsistencies between DPR, provincial DPRD and regency/municipality DPRD results of parties.
8 Karolin's result was third best among all elected MPs in Indonesia in the 2009 election. The highest number of votes went to Edhie Baskoro Yudhoyono, son of President Susilo Bambang Yudhoyono. The second highest result was that of Puan Maharani, daughter of Megawati Sukarnoputri and granddaughter of President Sukarno.

9 Except for the Kayong Utara regency, where PPD obtained 45% of the votes. PPD is consistently very strong in this *kabupaten*, with the *bupati* being from this party. Oesman Sapta from this party as a 2007 governor candidate also swept almost all votes in the sub-districts that later became Kayong Utara.

10 I used results for provincial assembly elections for this comparison to omit the discrepancy produced by the disproportionally high result for PDI-P's one candidate in the DPR election.

11 This was quite a similar situation to the 2004 presidential election when Kalla ran as Yudhoyono's running mate without his party's, Golkar's, endorsement (Mietzner 2013, chap. 5). Golkar's Aburizal Bakrie was touted to run as a presidential candidate since the party's decision in 2011, but Bakrie in the end failed to garner support from other parties. Bakrie was hoping until the last minute to run on Prabowo's ticket as his mate, but Prabowo opted for Hatta Rajasa. This left Golkar, the second strongest party in the parliament, out of the presidential race. Bakrie first decided to back Jokowi's ticket, but at the last minute changed his mind and endorsed Prabowo, causing a rift in the ranks of Golkar cadres (*Jakarta Post* 2014a, 2014b). Aspinall and Mietzner (2014) provide an excellent analysis of the background of the campaign and the campaign itself.

12 Sarumli Seneh, interview by the author, 13 September 2012.

13 Yakobus Kumis, interview by the author, 2 June 2011.

14 Sarumli Seneh, interview by the author, 13 September 2012.

15 Sarumli Seneh, interview by the author, 13 September 2012.

16 Sarumli Seneh, interview by the author, 13 September 2012.

17 Sunandar, interview by the author, 13 September 2012.

18 Davidson, however, refers to Yansen as a "local businessman of Chinese descent", not mentioning mixed ancestry (2008, 171).

19 As of May 2015, there has been little progress in establishing the Ketungau regency.

20 Except for Belitang Hulu, where Masiun's result was slightly higher than Abun's. However, Simon Petrus obtained 80% of the votes there, rendering the other candidates entirely irrelevant.

References

Antara. 2012. "Armyn-Fathan Sudah Mengundurkan Diri Dari Jabatan Negeri". *Antaranews.com Online*. www.antaranews.com

Aspinall, Edward, and Marcus Mietzner. 2014. "Indonesian Politics in 2014: Democracy's Close Call". *Bulletin of Indonesian Economic Studies* 50 (3): 347–369.

Badan Pusat Statistik. 2010. *Penduduk Menurut Wilayah dan Agama yang Dianut. Provinsi Kalimantan Barat*. Badan Pusat Statistik Online. www.sp2010.bps.go.id.

CAIREAU Center for Acceleration of Inter-Religious and Ethnic Understanding, ed. 2008. "Ethnic Dimension in Political Life: Study on Ethnic Preference in Pontianak Mayor Election". N.p. Unpublished.

Eilenberg, Michael. 2012. *At the Edges of States: Dynamics of State Formation in the Indonesian Borderlands*. Leiden: KITLV.

F. Alkap Pasti. 2003. "Dayak Islam Di Kalimantan Barat: Masa Lalu Dan Identitas Kini". In *Identitas Dan Postkolonialitas Di Indonesia*, edited by A. Budi Susanto, 105–142. Yogyakarta: Kanisius.

Harian Equator 2007. "Retno: Masih Muda, Punya Spirit Bangun Singkawang". 5 November.

———. 2010a. "Morkes Effendi Bertitah *Melayu Bersatu*". 10 May.

———. 2010b. "Mohon Dukungan Hingga di Pemerintahan". 14 May.

———. 2010c. "Dukungan Pasangan Yasir-Martin Terus Mengalir". 15 April.

———. 2010d. "Louren Optimis Yasyir-Martin Perhatikan Daerah Pedalaman". 23 April.

Hui, Yew-Foong. 2011. *Strangers at Home: History and Subjectivity among the Chinese Communities of West Kalimantan, Indonesia*. Leiden and Boston: BRILL.

Jakarta Post. 2014a. "Jokowi wins Golkar's support". 14 May.

———. 2014b "Golkar warns internal Jokowi-Kalla supporters to toe line". 20 May.

Kalimantan-news. 2011. "Kebakaran Lahan Di Kayan Hilir". *Kalimantan-news Online*. www.kalimantan-news.com.

———. 2012. "Mantan Bupati Sintang Jelaskan Proses Perjalanan Kabupaten Persiapan Ketungau". *Kalimantan-news Online*. www.kalimantan-news.com.

Kementerian Agama. 2010. "Jumlah Penduduk Menurut Agama per Kecamatan Propinsi Kalimantan Barat Tahun 2010". N.p. Unpublished.

Komisi Pemilihan Umum Kabupaten Kapuas Hulu. 2005. Rekapitulasi Perhitungan Suara Pemilihan Bupati Dan Wakil Bupati Tahun 2005 Tingkat Komisi Pemilihan Umum Kabupaten Kapuas Hulu. N.p.: Unpublished.

Komisi Pemilihan Umum Kabupaten Ketapang. 2005. Rekapitulasi Perhitungan Suara Pemilihan Bupati Dan Wakil Bupati Tahun 2005 Tingkat Komisi Pemilihan Umum Kabupaten Ketapang. N.p.: Unpublished.

———. 2010a. Rekapitulasi Perhitungan Suara Pemilihan Bupati Dan Wakil Bupati Tahun 2010 Tingkat Komisi Pemilihan Umum Kabupaten Ketapang. N.p.: Unpublished.

———. 2010b. Rekapitulasi Perhitungan Suara Putaran Kedua Pemilihan Bupati Dan Wakil Bupati Tahun 2010 Tingkat Komisi Pemilihan Umum Kabupaten Ketapang Putaran II. N.p.: Unpublished.

Komisi Pemilihan Umum Kabupaten Sanggau. 2008a. Rekapitulasi Perhitungan Suara Pemilihan Bupati Dan Wakil Bupati Tahun 2008 Tingkat Komisi Pemilihan Umum Kabupaten Sanggau Putaran Satu. N.p.: Unpublished.

———. 2008b. Rekapitulasi Perhitungan Suara Putaran II Pemilihan Kepala Daerah Dan Wakil Kepala Daerah Kabupaten Sanggau Tahun 2008. N.p.: Unpublished.

Komisi Pemilihan Umum Kabupaten Sekadau. 2005. Rekapitulasi Perhitungan Suara Pemilihan Bupati Dan Wakil Bupati Tahun 2005 Tingkat Komisi Pemilihan Umum Kabupaten Sekadau. N.p.: Unpublished.

Komisi Pemilihan Umum Kabupaten Sintang. 2006. Rekapitulasi Perhitungan Suara Pemilihan Bupati Dan Wakil Bupati Tahun 2006 Tingkat Komisi Pemilihan Umum Kabupaten Sintang. N.p.: Unpublished.

———. 2010. Rekapitulasi Perhitungan Suara Pemilihan Bupati Dan Wakil Bupati Tahun 2010 Tingkat Komisi Pemilihan Umum Kabupaten Sintang. N.p.: Unpublished.

Komisi Pemilihan Umum Kabupaten Kapuas Hulu. 2010. Rekapitulasi Perhitungan Suara Pemilihan Bupati Dan Wakil Bupati Tahun 2010 Tingkat Komisi Pemilihan Umum Kabupaten Kapuas Hulu. N.p.: Unpublished.

Komisi Pemilihan Umum Provinsi Kalimantan Barat. 2007. Perolehan Suara Sah Pemilu Gubernur Dan Wakil Gubernur Kalimantan Barat Tahun 2007 Per Kabupaten/Kota. N.p.: Unpublished.

———. 2012. Pengumuman Nomor 04/KPU-Prov-019-VII-2012 Tentang Nomor Urut Pasangan Calon Gubernur dan Wakil Gubernur dalam Pemilu Gubernur dan Wakil Gubernur Provinsi Kalimantan Barat Tahun 2012. Pontianak: Komisi Pemilihan Umum Provinsi Kalimantan Barat.

Komisi Pemilihan Umum Sekadau. 2010. Rekapitulasi Perhitungan Suara Pemilihan Bupati Dan Wakil Bupati Tahun 2010 Tingkat Komisi Pemilihan Umum Kabupaten Sekadau. N.p.: Unpublished.

Komisi Pemilihan Umum. 2004. Rekapitulasi Perhitungan Suara Pemilihan Umum Presiden dan Wakil Presiden Tingkat Komisi Pemilihan Umum Provinsi Kalimantan Barat; Putaran Kedua. Pontianak: Unpublished.

———. 2009a. Hasil Perolehan Suara Partai Politik Dewan Perwakilan Rakyat (DPR), Dewan Perwakilan Rakyat Daerah (DPRD) Tingkat Provinsi Kalimantan Barat Pemilu Tahun 2009. N.p.: Unpublished.

———. 2009b. Rekapitulasi Perhitungan Suara Pemilihan Umum Presiden dan Wakil Presiden Tingkat Komisi Pemilihan Umum Provinsi Kalimantan Barat. Pontianak: Unpublished.

———. 2011. Hasil Perolehan Suara Partai Politik Dewan Perwakilan Rakyat (DPR), Dewan Perwakilan Rakyat Daerah (DPRD) Tingkat Provinsi Kalimantan Barat Pemilu Tahun 2014. N.p.: Unpublished. Kompas, 2011. "Husin Salah Sebut Nama Partainya di Depan Prabowo". 22 November.

———. 2014. "Tiba di Pontianak, Jokowi Disambut Sorak Warga". 23 June.

Lingkaran Survei Indonesia. 2008. "Faktor Etnis Dalam Pilkada". *Kajian Bulanan* 09. www.lsi.co.id.

Maulana, Ardian, and Hokky Situngkir. 2009. "Coalitions in Multiparty System: Empirical Reflection of the Indonesian Regional Elections". http://ssrn.com/abstract=1400724

Merdeka.com. 2009. "Morkes Pimpin Golkar Kaltim". *Merdeka.com Online*. www.merderka.com.

Mietzner, Marcus. 2013. *Money, Power, and Ideology: Political Parties in Post-Authoritarian Indonesia*. Honolulu: University of Hawaii Press.

———. 2014. "How Jokowi Won and Democracy Survived". *Journal of Democracy* 25 (4): 111–125.

Pontianak Post. 2008. "Selebaran SARA ditemukan". 20 October.

———. 2009a. "Mega Ingin Ulangi Sukses di Kalbar". 26 March.

———. 2009b. "Mega Tekankan Pendidikan Berkelanjutan". 16 June.

———. 2010. "Ponpes Darun Nasyiin Dukung KH Suhri Maksudi". 22 October.

———. 2012a. "Cornelis-Christiandy Menang Mutlak". 27 September.

———. 2012b. "Milton Dipastikan Berlabuh di Golkar". 17 July.

———. 2012c. "Morkes-Burhan-Milton: Bersatu Wujudkan PKR". 11 September.

———. 2012d. "Petahana Bertahan". 26 September.

Rakyat Kalbar. 2012. Tambul: Hanya Orang Hulu yang Tahy Keinginan Masyarakat Timur Kalbar'. 10 September.

Subianto, Benny. 2009. "Ethnic Politics and the Rise of the Dayak Bureaucrats in Local Elections: Pilkada in Six Kabupaten in West Kalimantan". In *Deepening Democracy in Indonesia?: Direct Elections for Local Leaders (Pilkada)*, edited by Maribeth Erb and Priyambudi Sulistiyanto, 327–351. Singapore: ISEAS.

Tan, Paige Johnson. 2006. "Indonesia Seven Years after Soeharto: Party System Institutionalization in a New Democracy". *Contemporary Southeast Asia: A Journal of International and Strategic Affairs* 28 (1): 88–114.

Tanasaldy, Taufiq. 2007. "Ethnic Identity Politics in West Kalimantan". In *Local Politics in Post-Suharto Indonesia: Renegotiating Boundaries*, edited by Henk Schulte Nordholt and Gerry van Klinken. Leiden: KITLV.

———. 2012. *Regime Change and Ethnic Politics in Indonesia: Dayak Politics in West Kalimantan*. Leiden: KITLV.

Tribun Pontianak. 2012a. "Demokrat Kalbar Tawarkan Tiga Nama Cawagub ke Cornelis". 24 May.

———. 2012b. "Lima Partai Daftarkan Ria Norsan-Gusti Ramlana ke KPU". 18 June.

———. 2012c. "Cornelis: Kapuas Raya Salah Nama". 18 June.

———. 2014. "Hasil Lengkap Perolehan Suara Capres di Kalimantan Barat". 18 July.

Woodward, Mark. 2008. "Indonesia's Religious Political Parties: Democratic Consolidation and Security in Post-New Order Indonesia". *Asian Security* 4 (1): 41–60.

6 Sarawak and West Kalimantan

A comparison

The current research is based on some assumptions originating from classifying the studied cases as either consociational (Sarawak) or centripetal (West Kalimantan) institutional designs. Many questions that guided this research have been asked in studies conducted in other parts of the world, also in a comparative manner, but as Hicken pointed out, several of the Southeast Asian states could offer unique contributions to several ongoing debates in the socio-political literature that so far have been drawing from cases in other parts of the world. Hicken suggests that one such area that could benefit from Southeast Asian research is "parties and elections in divided societies" (2008, 92). This book took up Hicken's challenge: the in-depth analysis of ethno-politics in Malaysian Sarawak and Indonesian West Kalimantan was conducted in order to find dynamics of ethnic identity change under differing institutional settings. The study started off with hypotheses about political institutions' impact on ethnic identity change, and additional questions about the role of political parties and centre–periphery relations on ethno-politics. Later in this chapter I recapitulate the most important findings of this study in a comparative manner. I return to the main questions asked in Chapter 1 and answer them based on the analysis of each case; most importantly, I show how particular institutional solutions adopted in the analyzed polities influence the speed and frequency of ethnic identity change. I also indicate possible directions of further studies for each discussed element. The final paragraphs are policy recommendations.

6.1 Hypotheses

The first hypothesis said that fewer directly electable offices result in activation of fewer categories, while more direct elections induce a higher number of activated categories across the society. Sarawak's case showed that the first part of the hypothesis does not have to be true and is not true for Sarawak. Malaysia holds only two elections (to the state and national legislative assemblies), and indeed there was no difference found between categories activated in the two elections (state and national levels). Simply, the delineation of constituencies is such that voters find themselves in the same ethnic-majority constituency for both elections, and the candidates for both seats compete as members of the same category.

Therefore, there is no identity change from state election to national election or vice versa. Nevertheless, categories are activated through channels other than the electoral competition, and a variety of categories is activated during the elections. Most importantly, although titular identity of constituencies is fixed, the identities activated by parties are different from those activated at the level of constituency. Hence, despite this lack of identity change between elections at different levels, there were at least two and often three categories activated in politics (through parties and executive nominations) for each resident in Sarawak.

West Kalimantan showed that the second part of the hypothesis is also not necessarily true. The presidential election, analyzed from the provincial perspective, showed that the ethnic cleavage of this election is a replication of a cleavage also dominant in other spheres of political life in the province, most notably, in the gubernatorial elections. Hence, the presidential election does not add any alternative categories to the set of activated identities in West Kalimantan. Most categories alternative to the predominant ones were activated at the third tier (regency/city) of executive elections, although not in all units. Here Catholics were shown to compete against Protestants, or linguistic categories within the Dayaks were organizing politically to compete for the regent's position. The second direct gubernatorial election also showed differing preferences of Malays depending on their region of origin in the province. The legislative elections to the national-level assembly in Indonesia were also dominated by the same cleavage as the gubernatorial election. Because few categories are activated in politics, and in many regions individuals continue to retain only one category in their repertoires, no shifts between categories and dimensions were observed. In other regions, where the district-level elections led to activation of locally relevant categories, individuals were shown to retain two or three categories in their repertoires, but shifts between them were less frequent than in the case of Sarawak. One notable exception were cases in which a candidate mobilized an ethnic constituency (e.g. Muslim) in one election or round of election, but was able to attract entirely non-ethnic support in another round or election.

The second hypothesis referred to implicit and explicit forms of ethnic mobilization, and this hypothesis was confirmed in this research. Implicit activation of ethnic identities was linked to the slower pace of the process, and the likelihood of merging categories of different dimensions (e.g. religion and race). In Indonesia, where activation is implicit except for religious categories, I showed that category "Dayak" is activated partly through invoking the category "Christian", and distinction between the two is often impossible – whether for a researcher, or, I argue, a voter. Implicit mobilization involves creating alloys of partially overlapping categories (e.g. "Dayak" and "Christian"): the governor of West Kalimantan is a "Christian Dayak" all the time. He does not shift between these categories, being Christian on one occasion and Dayak on another. He is never referred to in these terms and denies representing any of these categories in public. In the same way, his electorate is not induced to identify once as "Christian" and once as "Dayak", but is mobilized to behave as "Christian Dayaks". However, as a regent candidate, he enjoyed greater support from his Kenayatn co-ethnics than from other Dayak categories in the regency – clearly Kenayatn mattered at the *kabupaten* level, but was of no importance on the provincial level. While these are logical and well-studied phenomena, it is important to have them in mind when discussing the problem of implicit and explicit ethnic mobilization. Individual candidates were shown to have little problem reaching out to different ethnic (or ethnically neutral) electoral groups, whether in an implicit or explicit manner. A further qualitative study of candidates' ethnic strategies and tools of mobilization in individual cases, to establish shifts of the activation strategies, should be a logical consequence of this research.

Explicit ethnic mobilization, observed in Malaysia and in its state of Sarawak, accounted for frequent and speedy identity changes and individuals retaining several activated identities in their repertoires. Verbally naming a category, without the necessity to resort to as sophisticated elements as ethnic outfits, or as imprecise message conveyors as family name, proved a powerful means to mobilize identities. A voter can identify as "Dayak" through a party, as a "Bidayuh" when supporting a Bidayuh ministerial nomination, and vote for a "Salako-speaking Bidayuh" in a state legislative election. All these categories are present in public discourse and are potentially included in elite bargaining. Explicit mobilization, however, was of enormous advantage to parties, as the Sarawak example showed; conversely, Indonesian parties in West Kalimantan maintained an implicitly invoked ethnic image that was relatively steady over time and locality. As I discuss later in this chapter, however, the "ethnic" property of parties in Indonesia must be seen as a strictly local phenomenon. One party may represent entirely different ethnic categories across the country and be strictly non-ethnic nationwide. Theoretical implications of this claim deserve further investigation, possibly across the country and in comparison with potential similar parties elsewhere.

The third hypothesis was about consociational polity, which, I hypothesized, would arrest ethnic identities by perpetuating activation of the same categories and dimensions over time and elections. We saw that in Sarawak, categories were not arrested over time, and political parties and executive nominations were able to induce ethnic identity change that made up for the relative fixedness of categories mobilized at the level of legislative constituencies. Sarawak partly retained the ethnic categories that were activated prior to the creation of the Malaysian Federation, but also a new ethnic split became relevant according to Malaysia-wide political practice, and the two dimensions (with some changes over time) are present in political dealings of the state at all times. On the other hand, the paradigm of elite bargaining as an approach to ethnic politics became entrenched in Sarawak. Ethnic coalitions came out as the established mind-set of most political entrepreneurs – even those who would otherwise aim to do away with political bargaining in non-ethnic aspects of social life. Therefore, although categories are not fixed and are not forcibly singular, we saw that there are at least two sets of categories along which the power is shared, which in itself is a valuable finding for a consociational polity – the modes of their activation are fixed. I argue that even the ultimate change of the ruling coalition and the opposition taking over power in Malaysia would not likely change this situation. The shape of the constituencies, drawn along ethnic lines, and the first-past-the-post electoral system will help perpetuate the ethnic composition of the legislative bodies and, by extension, will ensure that the paradigm of ethnic power sharing continues.

Surprisingly, however, the expectation that free ethnic competition induces frequent activation of different categories and enables shifts between identity dimensions was not entirely confirmed. Indonesian West Kalimantan, although a textbook example of multiple elections on three tiers of administration, provided more than enough examples of inertia in category activation. As it turned out, the presidential election was hijacked to reproduce the traditional Muslim versus non-Muslim division in West Kalimantan. Regencies where Muslims and Christians are more or less evenly numbered also saw a repetition of the religion-cleaved preferences. However, many *kabupaten* were also theatres of Protestant versus Catholic, North versus South and Dayak sub-groups' competitions. None of them, however, was able to trump the Malay versus Dayak or Muslim versus Christian one.

The most valuable find, however, came from the realization that both political systems involve – likely against the systems' designers' intentions – both power-sharing elements and free ethnic competition. In Malaysia, it was the political parties that in the quest for expanding their support sought to transgress the ethnic boundaries that the bargaining agreement would like them to uphold. Parties, both within the ruling coalition and the opposition, were finding ways of tricking the system and increasing their shares of power, but without upsetting the ethnic balance. In West Kalimantan, on the other hand, several informal practices were found that served to impose a non-institutionalized form of power sharing. While the dual ticket in executive elections gives the necessary institutional framework for it, the candidate selection usually took advantage of it to introduce an ethnic maximum-winning coalition. In other words, most regents shared power with vice-regents of an ethnic category that would otherwise be considered a political enemy to their own ethnic category. This is not to say that there is no electoral competition between these two categories, e.g. "Dayak" and "Malay". They compete fiercely, but there seems to be a tacit understanding that spoils are better shared, and hence vice-regents are picked from the contenders' side.

Let us look at particular institutional elements of the two political systems to see which of them contributed to more frequent identity change, and which were responsible for perpetuating the same set of categories over time.

1) Direct local and regional executive elections are opportunities for identity activation of categories different to those activated in legislative elections. However, if they take place within a constituency of the same composition as the legislative elections and political parties nominate candidates and run the campaigns, the relevant categories in the executive elections may still be the same as in the legislative elections. Nevertheless, several examples were shown of direct executive elections leading to ethnic identity change in Indonesia. If Malaysia was to introduce direct elections for division heads, given that the legislative constituencies are of different shape and ethnic composition than divisions, we might see entirely different categories activated in Sarawak than what we see under the current system.

Dual tickets in direct executive elections, combined with a limit of terms for officeholders, often lead to arresting of one set of categories and are not an incentive to activate new categories from one election to the next, as many regencies in Indonesia showed. Two ethnic categories (in particular if each is a large plurality), it turns out, switch at the position of power holders from term to term.

2) Executive members from nomination are poor tools of ethnic mobilization, especially if ministers come from the ranks of the legislative representatives. At the same time, Sarawak showed that these nominations are a channel of identity activation in some cases and are part of the delicate balance between different categories entitled to share power. This, however, is conditional on the explicit presence of ethnicity in politics. If naming of categories in which a minister is a member was forbidden or not part of political practice (see Indonesia), these nominations would have no or negligible bearing on identity activation.

3) Political parties contribute to identity change if explicit mobilization is allowed, and Sarawak offered multiple examples of it. Parties openly search for ways of expanding or consolidating their voters' base by discussing ethnic supporters they can gain in specific geographic areas. If mobilization is implicit and parties are not permitted to officially represent specific ethnic groups, they are likely to mobilize categories already activated in that polity, as it happens in West Kalimantan.

4) Creation of new legislative constituencies creates new ethnic majorities and is an effective way of activating new categories. This effect is strengthened if the constituencies can be created with a bottom-up impetus, e.g. like in the case of Indonesia, where localities can vie for a new regency or province. The new units will have an entirely new ethnic structure and very likely ethnic practice, or set of activated categories, and in many instances will have a new ethnic minimum-winning coalition. However, this mode of creating new constituencies is limited – new administrative units cannot be added endlessly. Therefore, after a period of reshaping of constituencies, the existing ones will become static and their ethnic composition will not change in the future. If the constituency creation and re-delineation is decided in a top-down manner (like in Sarawak), and the Election Commission decides on the shape of constituencies, the ethnic element can either be absent from considerations on the new shape of electoral districts, or may reflect interests of the ethnic category/party/coalition that holds power at the given time. Sarawak's case showed that some new constituencies offered the possibility of activating a new category, but these cases are rare and far apart. With more open and transparent involvement of different political groupings and the voters themselves, the institution of frequent constituency re-delineations known from Malaysia could be a ready tool of frequent identity changes.

5) The two cases analyzed here showed that federal/central intervention in regional/local ethnic politics happens chiefly through parties. The federal government in Kuala Lumpur was able to install the first Muslim Melanau as chief minister in Sarawak in 1970 mainly through its influence on the ruling coalition in Sarawak, and not through official, institutional channels. The understanding that only a Muslim can be the chief minister in Sarawak is decided at the coalition level

and although it is not formalized, it is invariably effective and strongly impacts ethnic relations in Sarawak. Arguably, mobilization of a common identity of all non-Muslim indigenous categories as one category fails, as potential gains from such mobilization are limited. Despite the numerical strength, the non-Muslim indigenous could not compete for the position of chief minister. In Indonesia, the central government seems too distant to be involved in ethnic politics of one of the country's 34 provinces. At the same time, central party leadership readily accepts the local ethnic outlook of PDI-P and participates in the ethnic venture of the West Kalimantanese chapter of the party. It is a mutually reinforcing strategy, in which a local ethnic category is mobilized around a party organization, which – albeit non-ethnic on the central level – offers supporting machinery for local elections. For the party, a loyal ethnic constituency in a province is invaluable as it guarantees strong electoral outcomes. In both cases, however, the central government's interventions (more direct in Sarawak and rather subtle in West Kalimantan) both reinforce existing ethnic practice and bolster long-activated categories, then induce identity change.

6.2   Parties and centre–periphery relations

Parties and centre–periphery relations were not hypothesized about in this research and were analyzed in an exploratory manner. They both were shown to be important intervening variables. Explicitly ethnic political parties in Sarawak were responsible for the activation of different ethnic categories to those activated via constituency majorities. Over time parties were also prone to look for new categories to appeal to and this way parties changed their ethnic outlook, becoming agents of ethnic identity change. In West Kalimantan, the important parties were only implicitly ethnic. Historically, the Dayaks as a category were shown to be able to seize a national party as their organization on the provincial level, and the current ethnic Dayak outlook of the PDI-P is a case in point. However, it requires time and sophistication to convey an ethnic image of a party through implicit means. It would be difficult to shift the ethnic appeal of an implicitly ethnic party from one election to the next, although the case of Sarawak (take SNAP in the 1970s) showed that it is possible if explicit ethnic appeal is allowed. Therefore, the implicit nature of parties' ethnic mobilization may result in conserving a party's ethnic outlook over time. Subsequent elections in Indonesia should help elaborate on this point.

Therefore, ethnic categories' activation success rates in Sarawak and West Kalimantan were conditional on both party cohesion and support from the national centre, which was channelled through the respective party in both cases. The institutional setting in each of the cases does not give the centre any particular privilege to intervene in ethnic dealings on the state or provincial level: Kuala Lumpur can only intervene in Sarawak power-sharing arrangements through the structures of Barisan Nasional. In Indonesia, the central government has no say in ethno-politics on the provincial level. Nevertheless, central party support is sought after by political entrepreneurs who wish to capitalize on ethnic identity in West Kalimantan, and parties do become channels of identity activation. As parties are also interested in maintaining a steady and loyal support in provinces, they readily welcome locally ethnic outlooks that their organizations maintain on the provincial level. This mutual interest between party organizations and ethnic categories ultimately makes ethnic identity activation dependent on support from the national centre and central elites gain the opportunity to impact ethno-politics on the sub-national level. This is the case of PDI-P and Dayaks. The Dayak category profits from the strong, well-rooted party organization, while PDI-P enjoys loyal support in the province and a ready pool of leaders to fill elected offices in the province and its regencies. This finding in West Kalimantan is entirely consistent with the result of analysis of intra-Barisan Nasional dynamics, in particular between the state chapter and the national chapter of the organization.

The most interesting empirical findings of this research revolve around the Muslim versus non-Muslim relations in the two analyzed units. While in Sarawak, the Muslim Malay/Melanau category has succeeded in securing power (after a short period of non-Muslim indigenous leadership), in West Kalimantan during Reformasi, non-Muslim Dayaks seized the power. Political parties, I argue, along with influences from the national centre, were responsible to a great extent for this outcome. In Sarawak, the party that represents the Malay/Melanau category (which is entirely Muslim, and almost all Muslims belong to this category), PBB, was historically the first to consolidate, and is the most coherent of all parties in Sarawak. Other parties either frequently changed their ethnic outlook, or struggled to secure the support of the majority of the category they claimed to represent. In West Kalimantan, on the other hand, the Dayak category (mostly Christians) is the only category which is identified with a particular party, PDI-P, and the party is specifically associated with Dayaks. This way, the electoral performance of an ethnic category is tied to the electoral performance of a party, and not only to particular candidates from within the category.

Both studied cases therefore confirmed the power of party organizations in securing ethnic identity activation and its maintenance over time. A category that has the institutional organization behind it was shown to be more likely to seize power and win over ethnic coalition partners. This was the case of the Malay/Melanau category in Sarawak in 1970. The Chinese could have entered a coalition with other parties, but they were drawn to the privileged position of the Malay/Melanau. The evidence in the case of West Kalimantan is much weaker and the observation time is much shorter; however, here the Chinese elite have entered a coalition with Dayaks and the Chinese vice-governor plays a role similar to that of the Chinese deputy chief ministers in Sarawak. Further studies of elections in Indonesia and West Kalimantan should show how long-lived the current coalition is and whether an alternative minimum-winning coalition (e.g. Dayak-Madurese or Malay-Javanese) can appear.

Another element that led the said categories to seize power is support received from the national centre. This is just too obvious in the case of the Malay/Melanau category in Sarawak, as Chapter 2 showed. Dayaks in West Kalimantan hardly enjoy the same privileges as Malays and Melanaus in Malaysia, but nevertheless, they enjoy support from Jakarta that no other category in the province does. While PDI-P needs the Dayak votes in the province, the Dayaks need direct access to the highest echelons of power in Jakarta. These mutual interests reinforce the connection between the party and the ethnic category.

Importantly, I found that in both countries studied that manipulation of size and shape of constituencies was the method of choice of political entrepreneurs to induce a desired ethnic outcome of elections. In Sarawak, constituency re-delineation is a routine exercise taking place every eight years. However, these changes impact only the ethnic composition of the legislative assembly, and each time only a limited number of constituencies are affected. In Indonesia, the creation of new administrative units, which in turn become new constituencies, is a much more difficult and time-consuming process, but the ethnic identity change induced in the process is incomparably more significant. New regencies and provinces create not only new proportions of ethnic categories (critical for electoral outcomes), but also an entirely new set of executive and legislative positions to be filled. The ethnic logic of re-delineation exercises in Sarawak (i.e. strengthening of one category's political influence) has been all too obvious and studying the process under the current regime can hardly be informative. However, if the government in Sarawak and Malaysia were to change, re-delineation exercises and their ethnic outcomes should be watched with utmost interest. A detailed study of elite ethnic bargains happening in the back stages of formation of new administrative units in West Kalimantan would also be invaluable for our understanding of ethno-politics in the province.

One of the empirically relevant questions was: *which* categories are activated in Sarawak and West Kalimantan, and *how* were they activated? I showed that in Sarawak the main sub-division, beyond the tripartite split (religion combined with race), was geographically induced. "Bidayuh" became a category of speakers of four different languages, who were non-Muslims and, most importantly, lived in inland areas of the western parts of Sarawak. Similarly, "Orang Ulu" became a category not because of a skilful leader (as in the case of "Malay/Melanau") or a political party (as in the case of "Iban"), but because of constituencies created in the area where an otherwise diverse population lived. As power was shared between categories that are assigned a number of legislators (in Sarawak the tokens of power sharing are representatives in legislative assemblies), the "Orang Ulu" category was assigned constituencies and legislators, and "Orang Ulu" as a category became a stakeholder in power in Sarawak. The primary division between the two Iban categories ("Rajang" and "Saribas") originated not only from the geography of their distribution, but also from party mobilization. Party activities were also behind the activation of two Chinese categories, which now are of lesser importance, also because of party politics.

West Kalimantanese categories, beyond the classic tripartite division, were shown to be split along two dimensions. One of them was again geography, which accounts for the differing preferences of "Pontianak Malays", "Sambas Malays" and other "Malays". This division was, however, so far only visible in one election (2012 gubernatorial election). Within the "Dayak" category, geographically induced activation of categories was present in selected regencies (Sintang, Seka-dau, Melawi); in other instances, the most prominent division within the Dayak category was Catholic versus Protestant. Hence, size, shape and geographic loca-tion of constituencies were the primary decisive factors inducing the activation of particular categories in Sarawak, and to an extent in West Kalimantan. Avail-ability of explicit information about candidates was another critical element that influenced the activation of sub-Dayak categories.

This study was yet another one attempting at capturing the identity change, and aiming not only to prove that identities are in flux, but also that their change can be traced back to political factors, e.g. institutions. One needs to answer two ques-tions after embarking on such an ambitious project. The first one is related to the academic value of this venture. Can evidence strong enough be provided to link the identity change to the factors we expect to influence it? The underlying doubt lies in the potential of identities simply to be fluid and impossible to capture. If the voter claims membership in four or five different categories, and each of the candidates can represent several categories, how are we to know which of the cat-egories was activated? Was it "Malay" or was it "Muslim"? Was it "Protestant" or was it "from the eastern region"? Qualitative data poorly answer these questions. Linking the institutional framework with strategies of parties or candidates was relatively straightforward, but capturing the specific ethnic categories that were activated in the process was much more difficult.

Moreover, in order to analyze ethnic identity change, one needs a plethora of data: not only must the census collect the information, it also has to tabulate it correctly and the published results must correspond to the units that commonsen-sically allow the analysis of elections. Indonesia would otherwise be a paragon of data availability, given that the electoral units are coterminous with administrative units. Unfortunately, not only are the census data not published for the small-est (*kecamatan*) units, but in addition several tabulation mistakes reduce the data quality, especially on the micro level. With the religious information being the most available, this research fell into the trap of focussing on the forever dominant religious division. Without doubt, the religious cleavage is extremely strong in the analyzed province, but the evidence presented is so skewed towards the religious dimension as the data was the most easily available: statistical information all the way down to the *kecamatan* was provided; the Election Commission shared candidates' religious background on information pamphlets; candidates readily provided the said information.

These shortcomings are likely to plague other studies that aim at identifying ethnic categories that go beyond the census-established ones. Without proper data referring to these categories being collected or published, and explicit mobiliza-tion precluding candidates from naming the specific category they aim to activate, a rigid analysis becomes difficult. Comparisons across countries characterized by different paradigms of accommodating ethnicity in politics are bound to be meth-odologically troubled, as centripetalist institutions aim to eschew ethnic references, which is likely to limit access to certain data and affect the way ethnic mobiliza-tion presents in electoral campaigns. Consociational polities will likely present a different trap: data will be collected on some ethnic categories but not others, and some categories will be explicitly discussed but not others, although there well may be attempts to mobilize other categories. This analysis, to an extent, was marred by both problems. In some cases in Indonesia, I was able to show regional preferences for candidates which imply ethnic mobilization, but was unable to name the category that was being mobilized. In other cases, I was able to appreciate efforts to mobilize a specific category (e.g. Madurese in West Kalimantan), but due to a lack of data I was unable to analyze how effective these efforts were.

Findings of this work can be used, to an extent, to assess the impact of the institutions on channelling ethnic competition in politics. The aforementioned distinction between explicit and implicit means of mobilization showed an interesting trait. Explicit mobilization, I discovered, allows for faster and more frequent changes of ethnic identity. It simply is easier to activate a category, if the category can be explicitly named. The Indonesian disinclination to explicit mobilization has the effect of driving the political entrepreneurs towards the one dimension of ethnicity that is legally approved – religion, or combining religion with other dimensions (culture or race), to make the activated categories alloys of several categories. Arguably, also without the privileged position of religion among other identities, the enforced implicit mobilization would have the same effect of arresting identities around those that had been activated during an earlier historical period – in this case, the previous political regime.

6.3 Theoretical findings

The theoretical framework applied in this research opened new venues for analysis of the nexus of ethnic identity and politics. The concept of "ethnic category" was used instead of "ethnic group", in order to accommodate the theoretical proposition that ethnic identities are multiple for each individual and a person belongs to several ethnic categories (and not to one ethnic group). The theoretical propositions required capturing of the ethnic identity in its continuous change. The dynamic of ethnic identity changes in the society therefore became the centre of the research. I looked for incentives of the change in political institutions.

The dynamic approach to ethnic identity in the political context was extremely rewarding. Especially Chapter 3 about Sarawak proves that the new dynamic approach yielded interesting results, even if most of the empirical data on which the analysis was based had been already known. By deploying political institutions as explanatory variables for ethnic identity change, I was able to propose a coherent timeline for political developments in Sarawak and offer a consistent elucidation of ethnic power relations in the state. Static analyses resting on a fixed set of categories (e.g. Dayak, Malay, Chinese) had been unable to show the changing party appeals, coalition partners' strengths and the ethnic elements of coalition-opposition rivalry. In the case of West Kalimantan, the application of the dynamic approach offered a chance to look at the very in flux current political situation. Assuming the ethnic categories activated during the Old and New Orders to be products of their times, I proposed that under a liberal democratic regime new categories would be activated. No such proposition would be possible if I had continued analyzing "ethnic groups".

The definition of ethnic parties deployed in this work, which is sensitive to time and place (i.e. party can be locally and temporarily ethnic), was shown to be a very powerful tool of analysis. Parties studied over time (especially in Sarawak) and between areas (especially in Indonesia) proved very susceptible to change. However, although I identified several parties as ethnic, I would not have been able to do so if I rigidly followed the condition of exclusiveness. Except for PBB (party excluding the non-Bumiputera) in Sarawak and some Muslim parties in Indonesia (all of which are of minor importance in West Kalimantan), parties analyzed in this research were not exclusive. Yet several of them I deemed ethnic at least at some point in time or in a particular area. Political practice in both studied countries discourages exclusiveness; nevertheless, parties do appeal to certain ethnic categories more than they do to others. Hence, although the condition of exclusiveness makes the ethnic party definition Chandra proposed rigid and clear-cut, it renders many parties non-ethnic, although the parties do fulfil most other conditions of ethnic parties. Moreover, based on parties analyzed in this study I argue that in the case of locally and temporarily ethnic parties, the condition of exclusiveness must be relaxed. A party ethnic in one area and non-ethnic in another (or everywhere else) is unlikely to convey any message tantamount to exclusiveness. Moreover, I would argue that explicit statements to this effect are ruled out altogether in the case of locally ethnic parties. Based on this study, insisting on the condition of exclusiveness would be counterproductive in the case of locally and temporarily ethnic parties.

Against expectations, it was found that a consociational polity does not have to rest on only one set of activated categories for its demography. Although Sarawak was found to have a power-sharing agreement and certain categories were obviously continuously activated to participate in that agreement, this polity was also shown to have several other categories politically activated, in different contexts. What I called *titular categories* were those that were entitled to a certain number of seats in the legislative elections and were represented by political parties. Other categories were activated if they happened to constitute a majority in a given constituency, and although that category would not be entitled to share power in the long term, in this election and in the particular constituency it was an activated category. Although the concept of titular categories can only be suitable for analysis of consociational arrangements, it comes in handy when we realize that a consociational polity may have other activated categories from the ones that share power according to institutionalized solutions.

An important observation related to the speed of ethnic identity change lies in the fact that all activated ethnic identity categories discussed in this book had an organizational structure behind them, or a definable geographic area belonging to them. All of the studied categories were activated through an ethnic organization (e.g. *adat* councils in West Kalimantan), political party, religious membership (also through church organizations, e.g. the Catholic church, or at least through explicit mention of the religious affiliation in an official setting) or electoral constituency with a particular category majority (see "Orang Ulu" in Sarawak, also combined with the establishment of the respective "Orang Ulu" association, or the proposed Ketungau regency seeking separation from Sintang in West Kalimantan). Institutional incentives, I discovered, were not enough to mobilize categories that were not already organized or explicitly existing in the political scene, or concentrated in areas that on their own could be separate constituencies.

Many observations in this book should provoke further research, which could be directed at four areas. The most obvious is to expand this study onto other countries. As these two cases showed, tracing ethnic identity change is a valuable way of uncovering hitherto unknown political dynamics, and some phenomena can be explained only when the link between ethnicity and politics is presented as politics' impact on ethnicity, and not as ethnicity's impact on politics. Therefore, further case studies similar to these ones are due. Similarly, a repeated study on Sarawak and West Kalimantan after a couple of decades could be equally informative. Both units are expected to be highly volatile politically – the electoral loss of Barisan Nasional in Sarawak or the provincial division of West Kalimantan would produce a new environment but with similar institutions – simply, a new laboratory for social experiments.

It is also worth looking at locally ethnic parties, their position in the party system and an ethnic party definition that can accommodate the phenomenon. In Sarawak, a good example of such a party was SUPP, which was a Chinese party in general (although not uniformly over time), but represented the interests of non-Muslim indigenous communities in constituencies in which it fielded candidates. In West Kalimantan, PDI-P was a strong example of a locally ethnic party and similar parties in other countries should be compared to ascertain their position in a party system, behaviour towards other parties and compatibility with non-ethnic or policy-based parties.

Finally, further research is required to establish whether the compatibility of power-sharing solutions and competitive ethnic politics uncovered here is specific for these two cases or whether this phenomenon may be more widespread or common. While neither of the two polities has institutions that specifically lead to the coexistence of elite bargaining and competitive outcomes, both cases were shown to display elements of both approaches. Sarawak, designed as consociational, relied heavily on fixed power-sharing solutions, but there was a relatively wide margin for manoeuvre for those political entrepreneurs who wished to expand the political influence of their parties or ethnic categories. West Kalimantan is not supposed to have ethnic mobilization, and there are definitely no institutional arrangements to induce ethnic power sharing. At the same time, the process of allocating positions based on ethnicity started in the province with the fall of Suharto and found its way to the current political competition.

6.4 Policy recommendations

Both countries analyzed in this research deploy policies aimed at curbing ethnic competition in politics and each country does it by different means. I argued that Malaysian politics rests on consociational principles, of which elite ethnic bargaining is the main element, and the prominence of explicit ethnic categories is indispensable for the system to function. Moreover, Malaysia as a federal structure is, in theory, better equipped to accommodate ethnic specificity at the state level. The Indonesian political setting was shown to be centripetal in its outlook, with several means adopted to eliminate ethnicity from political life; the unitary (albeit, as of the past decade, decentralized) structure allows – again in theory – less space for manipulation of province-level ethnic divisions.

Somewhat astonishingly, it was found that also in West Kalimantan, on several occasions political entrepreneurs engaged in the type of negotiations and settlements that strictly resemble power sharing. It is a key finding if one keeps in mind that West Kalimantan has a long and recent history of violent ethnic tensions in which many ethnic groups were involved at some point in history. It appears that a society that experienced a deadly ethnic conflict in its most recent history is likely to opt for political tools that immediately and to everyone's accord guarantee each relevant segment of the society participation in power. At the beginning of Chapter 2, I presented many potential dilemmas related to establishing those "relevant segments" in terms of ethnic categories in the case of Sarawak, but West Kalimantan proves that violent conflict does indeed simplify the question as to which ethnic categories should share power, and – even more importantly – strengthens the perceived need to establish a fixed power-sharing scheme.

Although political scientists can present logical, insightful and theoretically convincing arguments against the adoption of a consociational structure for the long run, these arguments might not be well heard or understood in the environment of day-to-day politics, elections and post-conflict fears. It appears that an agreement between the hitherto fighting parties in the form of a power-sharing deal is a powerful message that speaks to voters and is a promise of a peaceful coexistence, and therefore might be a preferred solution for societies where a risk of conflict is imminent and long-term goals seem too intangible to be considered. One way of looking at it is that including all ethnic segments in the government according to fixed proportions is a quick, prominent, easy to achieve tool that can be implemented with almost immediate effect. It is also a solution that seems "fair" or "just", and with some effort, it may be presented as relatively transparent and achieved through a consensus. This gives consociationalism a strong upper hand as a proposition for societies with recent violent conflict experience. To see political leaders of the warring parties involved in a negotiation process that bears characteristics of fairness, transparency and consensus may be more viable than a centripetal solution that involves quite complex institutional arrangements, whose outcome may be impossible to project and advertise within the society.

The case of West Kalimantan shows that electoral institutions that might be designed to take ethnicity out of the political equation, and in this way secure lasting peace (like centripetalism), might not be understood as such by the participants of the political process, or the goal that these institutions are supposed to achieve might not be clearly visible to them. Institutional arrangements deployed to defuse ethnic mobilization do not easily translate into the vernacular understanding of what is fair or due. In centripetal settings, the ethnic results of the electoral competition are unknown prior to voting, and this alone may make voters and candidates wary of the process. Where democratic institutions are not well established – more often than not the case of societies plagued by violence – participants of the electoral process may have few reasons to trust that the elections will be fair and one ethnic category will not skew them to their own advantage.

Unlike power-sharing agreements that can be presented to the involved parties, including voters, through very crude terms, centripetal designs deploy more sophisticated tools than consociationalism and it might seem to the players that the outcomes of the process are less known or foreseeable than in power-sharing agreements. This was argued to be the reason West Kalimantanese politicians often opt for ethnically mixed tickets for executive offices. If two ethnically exclusive candidate pairs compete against each other, the chance of each group winning is 50%. If each of the pairs includes both rivalling ethnic categories, the loss is less palpable, as at least the running mate is from the losing category and not only remains close to the power source but also has a chance to run and win in the next election, having established his/her position and popularity.

Sarawak also sheds some light into this situation: precisely because power is shared between some categories at some levels, at other levels more adventurous players can choose to opt for solutions that eschew the spoils of power sharing and seek advantages that go beyond the assigned quota. This is a riskier behaviour in the case of electoral loss, but it does not greatly endanger the fairly stable balance of ethnic coexistence. The relative security of one's interests represented through one ethnic category in a fixed-asset type agreement, like consociationalism, allows voters, candidates and parties to safely try and activate other ethnic categories in political situations that offer more freedom of choice of identity, i.e. are not part of the elite bargaining set-up. This seems to act like a safety valve: even if a particular ethnic category comes to perceive that its share of power within certain institutions is not satisfying, it can attempt to engage in ethnic competition for spoils through another institution, possibly as a differently defined category, instead of questioning the entire rationale behind the power-sharing agreement.

Having in mind societies that have recently experienced violent conflict and are potentially at risk of further violence, political institutions that combine consociationalist and centripetalist tools seem to be of advantage. At some level of the political life, e.g. the executive, a power-sharing agreement should be encouraged, for the reasons described earlier: all participants of the political life appreciate the security of a pre-agreed bargain. At the same time, having the long-term effects in mind and aiming at diminishing the chances of reifying one set of ethnic categories, other measures can be taken to lessen the value of ethnic manipulation in other spheres of political life, e.g. through an ethnic party ban, or constituency boundaries that dissect ethnic regions.

In this light, the consociational element must be deemed entirely obsolete for societies that have no recent memory of conflict. In the light of the arguments presented here, there is no added value of elite bargaining in countries where the electoral process is trusted and political leaders have no history of instigating ethnic violence or spreading hatred. Although Sarawak's example shows that attempts at increasing one's share of power are frequent even in consociational arrangements, it is beyond doubt that in Sarawak the freedom of electoral choice for voters is severely curtailed and cannot be justified if the ultimate goal is a body of mature political participants who make choices based on policy platforms. As long as politics revolves around the current pattern of attempts to incrementally improve one's ethnic categories' representation in politics, a shift towards policy-based political choice is unlikely.

Having the normative idea in mind, that more activated identities are better, we should further look for ways of inducing identity activation. Multiple elections do have an impact on the number of categories activated, as the Indonesian case showed, but to a much more limited extent than expected, and conditionally on other factors. Regent elections in Indonesia do provide an opportunity to activate alternative ethnic categories to those that cleaved the gubernatorial elections in West Kalimantan. However, most regent elections remained chiefly Christians versus Muslims affairs, even if other secondary cleavages were present, especially that religious composition of many regencies renders Christians and Muslims minimum-winning coalitions on their own. Therefore, the ethnic identity activation is strongly dependent on the shape and ethnic proportions of constituencies. A similar trait was found in Sarawak; re-delineation of constituencies, we found, could result in the activation of a new category altogether. Therefore, the proposition should go: in order to maintain the high dynamic of ethnic identity change, the shape of constituencies should be frequently altered, in order to create new minimum-winning coalitions, and consequently induce the activation of new categories. Although the proposition is hardly feasible, the evidence strongly supports it. While entirely impossible for direct executive elections (here constituencies are coterminous with administrative units), a frequent and significant re-delineation of constituencies may resolve the issue of entrenched categories activated in legislative constituencies, as it is in Sarawak.

Reference

Hicken, Allen. 2008. "Developing Democracies in Southeast Asia: Theorizing the Role of Parties and Elections". In *Southeast Asia in Political Science: Theory, Region and Qualitative Analysis*, edited by Erik Martinez Kuhonta, Dan Slater and Tuong Vu, 80–101. Stanford, CA: Stanford University Press.

Appendices

1. Selected Malaysian party logos

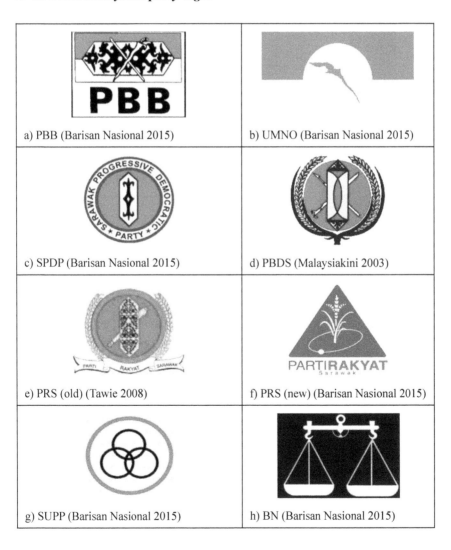

a) PBB (Barisan Nasional 2015)	b) UMNO (Barisan Nasional 2015)
c) SPDP (Barisan Nasional 2015)	d) PBDS (Malaysiakini 2003)
e) PRS (old) (Tawie 2008)	f) PRS (new) (Barisan Nasional 2015)
g) SUPP (Barisan Nasional 2015)	h) BN (Barisan Nasional 2015)

2. Election commission poster for the 2007 gubernatorial election, West Kalimantan

Source: KPU Kalbar (2007)

3. Election commission poster for the 2012 gubernatorial election, West Kalimantan

Source: KPU Kalbar (2012)

References

Barisan Nasional. 2015. 'Parti Komponen'. www.barisannasional.org.my.

KPU Kalbar. 2007. 'Pasangan Calon Pemilu Gubernur dan Wakil Gubernur Kalbar 2007'. Pontianak. Komisi Pemilihan Umum.

KPU Kalbar. 2012. 'Pengumuman nomor: 04-KPU-Prov-019-VIII-2012'. Pontianak: Komisi Pemilihan Umum.

Malaysiakini. 2003. 'A month to PBDS election, campaign gets hotter'. www.Malaysiakini.com.

Tawie, Joseph. 2008. 'PRS to submit memorandum on NCR land'. www.thebrokenshield.blogspot.com. 25 November 2008.

Index

Milton Keynes UK
Ingram Content Group UK Ltd.
UKHW031148141024
449569UK00024B/965